Computed Tomography of the Spine
Diagnostic Exercises

Computed Tomography of the Spine
Diagnostic Exercises

Robert Kricun, M.D.
Department of Radiology
Lehigh Valley Hospital Center
Allentown, Pennsylvania

Clinical Assistant Professor
Department of Radiology
Hospital University of Pennsylvania
Philadelphia, Pennsylvania

Morrie E. Kricun, M.D.
Associate Professor
Department of Radiology
Orthopedic Radiology Section
Hospital University of Pennsylvania
Philadelphia, Pennsylvania

AN ASPEN PUBLICATION®
Aspen Publishers, Inc.
1987
Rockville, Maryland
Royal Tunbridge Wells

Library of Congress Cataloging-in-Publication Data

Kricun, Robert.
Computed tomography of the spine.

"An Aspen publication."
Includes bibliographies and index.
1. Spine—Diseases—Diagnosis—Case studies. 2. Spine—Radiography—case studies. 3. Tomography—Case studies. I. Kricun, Morrison E., 1938-   . II. Title. [DNLM: 1. Spine—radiography—case studies. 2. Tomography, X-Ray Computed—case studies. WE 725 K92c]
RD768.K74   1987   617′.37507′572   86-28688
ISBN: 0-87189-375-4

Copyright © 1987 by Aspen Publishers, Inc.
All rights reserved.

Aspen Publishers, Inc. grants permission for photocopying for personal or internal use, or for the personal or internal use of specific clients registered with the Copyright Clearance Center (CCC). This consent is given on the condition that the copier pay a $1.00 fee plus $.12 per page for each photocopy through the CCC for photocopying beyond that permitted by the U.S. Copyright Law. The fee should be paid directly to the CCC, 21 Congress St., Salem, Massachusetts 01970.
0-87189-375-4/87 $1.00 + .12.

This consent does not extend to other kinds of copying, such as copying for general distribution, for advertising or promotional purposes, for creating new collective works, or for resale. For information, address Aspen Publishers, Inc., 1600 Research Boulevard, Rockville, Maryland 20850.

Editorial Services: Jane Coyle

Library of Congress Catalog Card Number: 86-28688
ISBN: 0-87189-375-4

Printed in the United States of America

1 2 3 4 5

To our mother
Esther G. Kricun
and to the memory of our father
Dr. A. Alfred Kricun

To Stephanie, Ashley, and Bret
R. K.

To Ginny
M. E. K.

FOREWORD

Since its introduction as a clinical tool for the evaluation of human disease little more than 12 years ago, computed tomography has evolved into one of the most reliable of diagnostic techniques. Its impact on health care is a matter of record and is nowhere more dramatic than in the study of the brain and spinal cord. Superior contrast resolution and a capacity to provide quantitative measurements of tissue are among the attributes of computed tomography that ensure its successful application to the diagnosis of spinal disorders. Like conventional tomography, computed tomography is planar but unlike the conventional technique, computed tomography provides transaxial images of the spine (and other sites) that sorely test the diagnostic acumen of the observer and, specifically, his or her knowledge of spinal anatomy. Doctors Robert and Morrie Kricun have recognized the unique challenge presented by this revolutionary imaging technique and have collaborated in the production of a book that addresses the issues that are fundamental to accurate computed tomographic diagnosis of spinal diseases. Both are well-known radiologists with a particular interest and expertise in such diseases, and the resulting book represents an important source of information to those of us with less expertise.

The format of the book is reminiscent of that used so successfully in the syllabi of the American College of Radiology, consisting of a series of more than 65 cases, each introduced by a challenge computed tomographic image or images and followed by an explanation of the findings, a thorough discussion of the specific diagnosis, and a concise list of important references that can be consulted for additional information. Anatomic drawings and photographs, routine radiographs, and myelograms are used to accent the computed tomographic displays and to provide perspective, and attention is directed to diagnostic contributions made by other imaging techniques, such as ultrasonography and magnetic resonance. Accessory computed tomography methods, including the use of reformatted images and intravenous contrast material, are illustrated.

The vertebral column is studied in its entirety, from the occiput above to the sacrum and sacroiliac joints below, with attention to abnormalities of both bone and soft tissue. Diseases of the vertebrae, intervertebral discs, spinal cord and its extensions, and connective tissue are addressed, including traumatic, neoplastic, infectious, congenital, degenerative, postoperative, and idopathic processes. In each instance, anatomic features pertinent to an understanding of the computed tomographic abnormalities are emphasized, terminology is precisely defined, and the relative contributions of each of the available imaging methods are indicated.

The authors are to be congratulated in providing a first-rate, up-to-date, readable account of the application of computed tomography to the evaluation of disorders of the vertebral column, and in doing so systematically, thoroughly, and in a fashion that ensures reader participation. The result is a timely and significant contribution to the medical literature, for which I am deeply honored to have been asked to prepare this Foreword.

Donald Resnick, M.D.
Professor of Radiology
University of California, San Diego

Chief, Radiology Service
Veterans Administration Medical Center
San Diego, California

PREFACE

Computed tomography (CT) has gained an increasingly important role in the evaluation of patients with spinal disorders. As the use of spinal CT has expanded, more physicians have become involved with this valuable imaging modality. We have prepared a text that we hope will benefit those interested in understanding and interpreting CT of the spine.

We have chosen a format that offers the reader an opportunity to examine the case materials of "unknowns" and to formulate a diagnostic opinion. On the page following each case presentation, the reader will find an in-depth discussion of the entity with special attention to the CT findings, use of CT in that particular disorder, and differential diagnosis. Included are discussions of CT technique and the use of intrathecal and intravenous contrast. We often present additional material to demonstrate the different appearances of the same disorder and the similar appearances of different entities. We have also used anatomic specimens and medical illustrations to aid the reader in better understanding the anatomy and pathology of the spine.

Although this book is presented as a series of unknowns, cases have been grouped by general topic such as disc disorders, spinal stenosis, infection, tumor, trauma, congenital disorders, the postoperative state, and so on. The reader should be cautioned, however, that "pitfalls" have been sprinkled throughout the book. Finally, up-to-date reference lists accompany the case material allowing the reader to continue with further study.

We believe that the format, case material, and discussions will provide the reader with diagnostic challenges and the knowledge needed to better utilize and interpret CT images of the spine.

R. K.
M. E. K.

ACKNOWLEDGMENTS

Preparation of this book required the time, effort, and talents of many people. The authors wish to acknowledge and thank those who have helped us so much.

First we thank Carol Gagnon, whose cheerful attitude and tireless effort made this project manageable and whose considerable talent enhanced its quality. Carol supplied not only her secretarial skills but all of the beautiful medical illustrations. Her work has added immeasurably to the quality of this book.

A book of this nature requires outstanding images and photography. We greatly appreciate the technical assistance of Eastman Kodak Co. with the radiographic reproductions. Ronald Pecoul was particularly helpful in this regard. Steve Strommer and G. Douglas Thayer provided additional photographic assistance for which we are grateful.

We thank our CT technologists Roseann Golba, Kathi McGill, Jane Miller, and Mary Joan Trembler for the care and concern they gave our patients and their CT examinations. In many instances they spent extra time rephotographing CT images to produce optimal examples. Alice Vrsan, the teaching file librarian, gathered valuable patient data.

The cooperation of many physicians enhanced this project. We thank Drs. Robert Zimmerman and Herbert Kressel for sharing magnetic resonance images with us. Dr. Spancer Borden IV provided us with pediatric case material. Dr. Zwu Shin Lin reviewed portions of our manuscript. Our many radiology colleagues gave needed support and our referring physicians provided important clinical correlations. We thank Dr. Donald Resnick for graciously contributing the foreword.

Ruby Richardson was very helpful and supportive when the project began at University Park Press. Later, Deborah Collins and Anne Patterson of Aspen Publishers made significant contributions and took the project to completion. We are also grateful to others on the staff at Aspen Publishers for their assistance.

We would like to thank the Dorothy Rider Pool Health Care Trust for generously funding this project.

Finally, we greatly appreciate the patience and encouragement of our families.

CASE 1

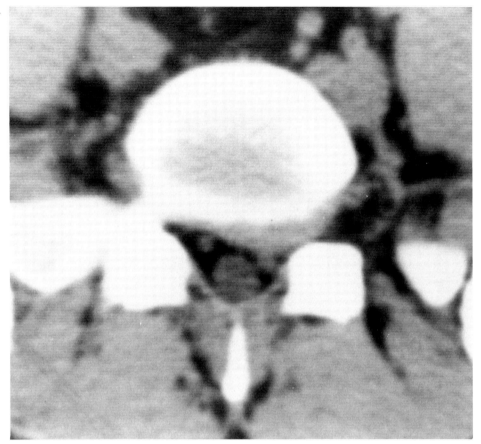

FIG. 1A. Axial CT at L5-S1. This 30-year-old male had back pain that radiated into the left leg.

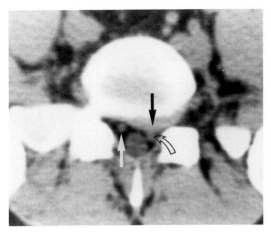

FIG. 1A-1. Posterolateral disc herniation. There is focal protrusion of the disc on the left *(black arrow)*, causing posterior displacement of the left S1 nerve root *(open arrow)* and obliteration of the anterior epidural fat. Compare with the normal position of the right S1 nerve root *(white arrow)* and presence of epidural fat on right.

Posterolateral Disc Herniation

There is a posterolateral herniation of the nucleus pulposus (HNP) at L5-S1 on the left side causing obliteration of the epidural fat and posterior displacement of the S1 nerve root (Fig. 1A-1). The disc is distinguished from the thecal sac by virtue of its higher CT density. Usually the disc measures in a range of 50 to 100 Hounsfield units (HU), whereas the thecal sac measures approximately 0 to 30 HU. The actual numerical measurements can be generated by the CT scanner; however, the density differences

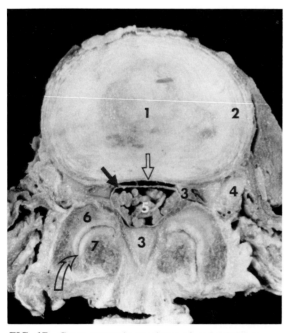

FIG. 1B. Gross anatomic specimen of an L4-L5 intervertebral disc in the axial plane. *1*, nucleus pulposus; *2*, annulus fibrosus; *3*, epidural fat; *4*, emerging L4 nerve; *5*, cauda equina within thecal sac; *6*, superior articular process of L5; *7*, inferior articular process of L4; *open straight arrow*, posterior longitudinal ligament; *closed arrow*, dura; *curved arrow*, facet joint.

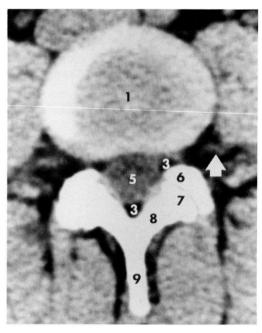

FIG. 1C. Axial CT at the normal L4-L5 intervertebral disc level. Notice the difference in CT densities between bone, disc, thecal sac, and fat. The posterior margin of the disc at this level is normally concave or flat. The nucleus pulposus and the annulus fibrosus cannot be differentiated. *1*, disc; *3*, epidural fat; *5*, thecal sac; *6*, superior articular process of L5; *7*, inferior articular process of L4; *8*, lamina of L4; *9*, spinous process of L4; *arrow*, emerging nerve of L4.

are usually apparent visually. Fat has negative CT attenuation values and is thus readily differentiated from the thecal sac. Appreciation of the differences in density between disc, thecal sac, and epidural fat is crucial in the CT evaluation of disc disease.

Let us begin the CT evaluation for disc herniation by carefully examining the posterior margin of the disc. At L5-S1, the posterior margin of the disc may be flat or slightly convex, whereas at L4-L5 and at the lumbar disc spaces above this level the posterior margin is flat or slightly concave.[8] A normal anatomic specimen and a comparable CT scan at L4-L5 are shown in Figs. 1B and 1C. Note the slightly concave posterior margin of the disc, the lack of compression of the thecal sac, and the preservation of epidural fat. The most important CT features of disc herniation are focal protrusion of the disc altering the normal configuration of the disc margin, displacement or compression of the nerve root, displacement or compression of the thecal sac, and displacement or obliteration of the epidural fat.[5,7,8,12,15] The major CT findings of HNP are depicted in Fig. 1D. At the L5-S1 level, the S1 nerve root is normally surrounded by abundant epidural fat and is readily visualized with CT (Fig. 1A-1). However, at the L4-L5 intervertebral disc level the L5 nerve root may not be visualized with CT. The diagnosis of posterolateral disc herniation is still made with confidence at this level (as well as at higher lumbar levels), although nerve root displacement may not be detected (Fig. 1E). The most common type of disc herniation is a "soft" herniation of the nucleus pulposus. A "hard" disc represents either calcified disc herniation or osteophyte formation and may cause compression of the thecal sac and nerve root similar to that seen with a "soft" or noncalcified disc herniation (Fig. 1F). As we will see in other cases, the posterior margin of the disc may be normal in cases of far lateral disc herniation (HNP into or beyond the neural foramen) and in cases in which an extruded disc is displaced cephalad or caudad to the intervertebral disc space level. Also, the posterior margin may appear normal when a free disc fragment is displaced posteriorly. The free fragment may present as a soft-tissue mass within the spinal canal separated from the disc margin by fat.

Over the years, myelography has been the gold standard for the diagnosis of disc disease. As more diagnosticians have gained experience with CT of the spine, the CT examination has achieved increasing acceptance as a primary study. Accordingly, we will

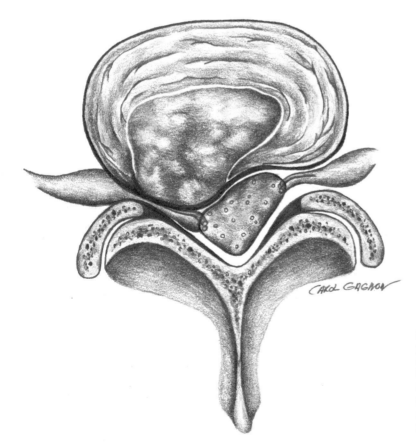

FIG 1D. Major characteristics of a posterolateral disc herniation. This drawing is a composite view of the intervertebral disc space and its adjacent levels. Posterolateral disc herniation compresses or displaces the nerve root, the epidural fat, and the thecal sac.

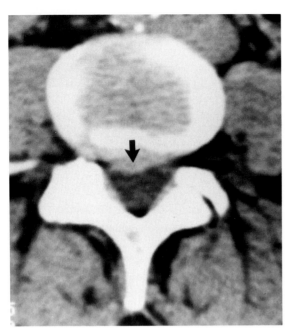 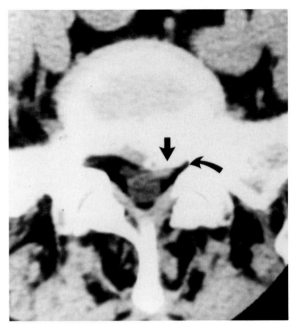

FIG. 1E. Posterolateral disc herniation. Axial CT at L4-L5. Focal protrusion of the disc to the right of the midline *(arrow)* is due to HNP. The difference in density between disc and thecal sac is readily apparent. Although the L5 nerve root may not be visualized with unenhanced CT, the diagnosis of disc herniation can still be made with confidence.

FIG. 1F. Hard disc. Axial CT at L5-S1. The "hard" disc *(straight arrow)* is distinguished from a typical "soft" disc herniation by its much denser appearance. It represents either a vertebral body osteophyte or a calcified disc and is causing compression of the left S1 nerve root, obliteration of the epidural fat, and narrowing of the lateral recess *(curved arrow)*.

attempt to correlate CT and myelography where it is instructive. CT and myelography are both accurate methods of diagnosing disc herniation in patients who have not had prior surgery; however, each modality has advantages and disadvantages. Although one report suggests that myelography with water-soluble contrast is significantly more accurate than CT in the diagnosis of disc herniation,[2] most investigators believe that CT compares favorably with myelography in the diagnosis of HNP.[1,3,5,9,12–15] In some series CT has been more accurate than myelography[8,9,13,14] primarily because of its superiority in diagnosing disc herniation at L5-S1 and in diagnosing lateral disc herniation. In several series CT has an overall accuracy of 92% to 94% in the diagnosis of HNP, with no significant difference at the various levels studied.[3,7,8,13,14] Myelography, on the other hand, is highly accurate at L3-L4 and L4-L5 but may be only 70% accurate at L5-S1.[8] Abundant epidural fat at L5-S1 may cause a wide separation of the disc from the thecal sac, leading to a false-negative myelogram. Unlike CT, which directly images disc herniation, the myelogram relies on indirect evidence of disc herniation such as angular deformity of the anterolateral thecal sac as well as widening or deformity of a nerve root[10] (Fig. 1G). Lateral disc herniation is another cause of a false-negative myelogram since myelography performed with water-soluble contrast demonstrates the nerve root sheath only as far as the dorsal root ganglion. Thus, a far lateral disc herniation into or lateral to the neural foramen may go undetected by myelography while being clearly demonstrated by CT.

Conventional CT (without intrathecal contrast) is usually sufficient and even desirable for evaluation of lumbar disc disease because of the natural density differences between the disc and the thecal sac (Figs. 1H, 1I). Occasionally, computed tomographic myelography (CTM) may provide additional information.[1,6] CTM may follow 2 to 6 hours after a myelogram performed with nonionic water-soluble metrizamide contrast (e.g., 15 mL of 190 mg of iodine per milliliter of metrizamide) or may follow immediately after the intrathecal introduction of low-dose contrast (e.g., 3 mL of 150 mg of iodine per milliliter of metrizamide). Newer nonionic water-soluble contrast agents (e.g., iopamidol and iohexol) have fewer and less severe side effects than metrizamide. CTM of the lumbar spine may follow 2 to 4 hours after a myelogram performed with 10 to 15 mL of 200 mg of iodine per milliliter of iopamidol or 10 to 15 mL of 180 mg of iodine per milliliter of iohexol.

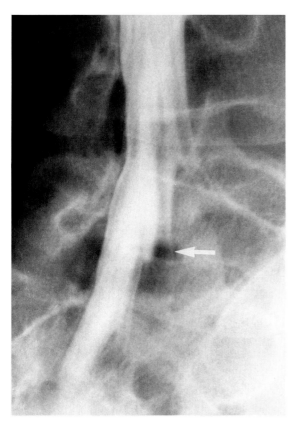

FIG. 1G. Posterolateral disc herniation. Oblique view from a lumbar myelogram performed with water-soluble contrast demonstrates abrupt termination and widening of the S1 nerve root sleeve *(arrow)*.

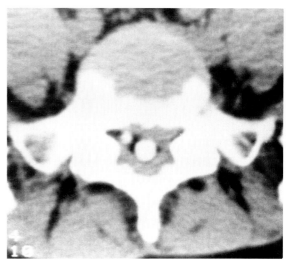

FIG. 1I. Posterolateral disc herniation. Same patient as in Figs. 1G and 1H. CTM followed a myelogram and is viewed at the same axial level as in Fig. 1H. The right S1 nerve root sleeve fills with contrast and is in the normal position. On the left side the disc herniation fills the anterior epidural space and prevents filling of the nerve root sleeve. This sequence of figures demonstrates the typical appearance of a posterolateral disc herniation, which can readily be diagnosed by myelography, CT, or CTM. The myelogram, CT, and CTM are shown in Figs. 1G, 1H, and 1I for their teaching value; however, the diagnosis of disc herniation does not require all three studies in this case or in most other cases.

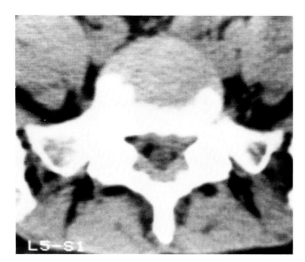

FIG. 1H. Posterolateral disc herniation. Same patient as in Fig. 1G. CT at L5-S1. The disc herniation has caused obliteration of the anterior epidural fat and compression of the S1 nerve root.

Low-dose CTM can be performed immediately after the introduction of 3 mL of iopamidol or iohexol. If these guidelines are used, the contrast agent should be visible but not too dense for optimal study. The patient should be turned from side to side immediately prior to the examination to prevent layering of the contrast posteriorly. If layering of the contrast occurs, scanning with the patient prone may be helpful.

An advantage of conventional myelography over CT is its ability to study all levels including the conus medullaris, whereas CT is usually but not always limited to three disc space levels. Thus a false-negative CT study occurs if the abnormal level is not examined. CT may also be falsely negative if the technique is inadequate (e.g., too few slices obtained) or the patient is too large for an optimal study. Occasionally, a massive disc herniation that fills the entire canal has been misdiagnosed as normal.[8,13]

Despite these limitations, CT is an accurate method of diagnosing HNP and is more accurate than myelography in its ability to diagnose lateral recess stenosis and neural foraminal stenosis, which not infrequently accompany disc herniation.[4] The preoperative diagnosis of lateral spinal stenosis accompanying disc herniation may lead to more appropriate surgical therapy and a subsequent decrease in the failed back surgery syndrome.[4] Some authors, citing diagnostic, psychological, and economic factors,[1,3,11] have concluded that CT should replace myelography as the initial imaging examination for evaluation of patients with low back pain. Although CT appears to be comparable to myelography in the diagnosis of HNP, it has the additional advantages of having less morbidity, producing less anxiety, and requiring no hospitalization. It is always important to compare the CT and myelographic findings with the patient's clinical symptoms and signs. However, when CT is used as the initial diagnostic modality it is imperative to compare the clinical and CT findings and to perform myelography if there is a lack of correlation.[3,9,11] In one study it was concluded that myelographic studies would be reduced fivefold if limited to cases in which the CT study was inconclusive.[12]

References

1. Anand AK, Lee BCP: Plain and metrizamide CT of lumbar disk disease: Comparison with myelography. *AJNR* 1982; 3:567–571.
2. Bell GR, Rothman RH, Booth RE, et al: A study of computer-assisted tomography: Comparison of metrizamide myelography and computed tomography in the diagnosis of herniated lumbar disc and spinal stenosis. *Spine* 1984;9:552–556.
3. Bosacco SJ, Berman AT, Garbarino JL, et al: A comparison of CT scanning and myelography in the diagnosis of lumbar disc herniation. *Clin Orthop* 1984;190:124–128.
4. Burton CV, Kirkaldy-Willis WH, Yong-Hing K, et al: Causes of failure of surgery on the lumbar spine. *Clin Orthop* 1981;157:191–199.
5. Carrera GF, Williams AL, Haughton VM: Computed tomography in sciatica. *Radiology* 1980;137:433–437.
6. Dublin AB, McGahan JP, Reid MH: The value of computed tomographic metrizamide myelography in the neuroradiological evaluation of the spine. *Radiology* 1983;146:79–86.
7. Firooznia H, Benjamin V, Kricheff II, et al: CT of lumbar spine disk herniation: Correlation with surgical findings. *AJNR* 1984;5:91–96, *AJR* 1984;142:587–592.
8. Fries JW, Abodeely DA, Vijungco JG, et al: Computed tomography of herniated and extruded nucleus pulposus. *J Comput Assist Tomogr* 1982;6:874–887.
9. Haughton VM, Eldevik OP, Magnaes B, et al: A prospective comparison of computed tomography and myelography in the diagnosis of herniated lumbar disks. *Radiology* 1982;142:103–110.
10. Kieffer SA, Sherry RG, Wellenstein DE, et al: Bulging lumbar intervertebral disk: Myelographic differentiation from herniated disk with nerve root compression. *AJR* 1982;138:709–716.
11. Mall JC, Kaiser JA: Critical evaluation of computed tomography versus myelography in assessing low back pain, in Genant HK (ed): *Spine Update 1984: Perspectives in Radiology, Orthopaedic Surgery, and Neurosurgery.* San Francisco, Radiology Research and Education Foundation, 1983, pp 97–105.
12. Raskin SP, Keating JW: Recognition of lumbar disk disease: Comparison of myelography and computed tomography. *AJR* 1982;139:349–355.
13. Tchang SPK, Howie JL, Kirkaldy-Willis WH, et al: Computed tomography versus myelography in diagnosis of lumbar disc herniation. *J Can Assoc Radiol* 1982;33:15–20.
14. Teplick JG, Haskin ME: CT and lumbar disc herniation. *Radiol Clin North Am* 1983;21:259–288.
15. Williams AL, Haughton VM, Syvertsen A: Computed tomography in the diagnosis of herniated nucleus pulposus. *Radiology* 1980;135:95–99.

CASE 2

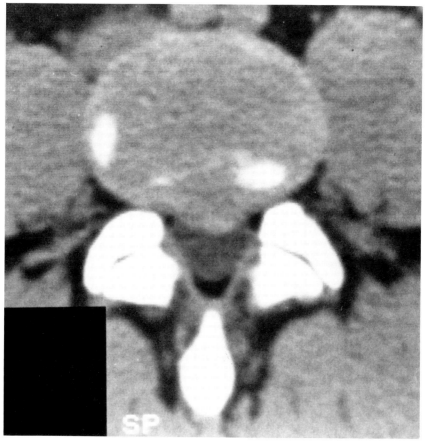

FIG. 2A. Axial CT scan at the L4-L5 intervertebral disc space obtained for evaluation of low back pain.

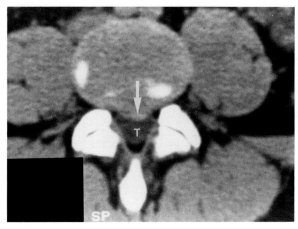

FIG. 2A-1. Central disc herniation. Focal midline protrusion of the disc posteriorly *(arrow)* causes moderate compression of the lower density thecal sac *(T)*.

Central Disc Herniation

There is alteration of the posterior margin of the L4-L5 disc with focal midline protrusion of the disc causing compression of the thecal sac (Fig. 2A-1). This is a subligamentous central disc herniation. Various subtle differences in the terminology describing disc disease have been forwarded in the literature. However, we use the terms bulging disc, herniated disc, extruded disc, and free disc fragment as they have been described previously by other authors.[1,4] A schematic representation of these disc disorders is shown in Fig. 2B. A bulging disc is a uniform, generalized protrusion of the annulus fibrosus beyond the vertebral body margin. A herniated disc is present when a portion of the nucleus pulposus ruptures through a tear in the annulus fibrosus. When the herniated nucleus pulposus remains confined beneath the posterior longitudinal ligament, it is considered a subligamentous herniation. When nuclear material traverses around or through the posterior longitudinal ligament, the disc is termed extruded. A free disc fragment is an extruded nucleus pulposus that lies free in the epidural space or, rarely, in the intradural compartment. Examples of the various manifestations of disc disease are presented in subsequent cases.

Disc herniations may be described as posterolateral (60% to 85%), central (5% to 35%), or lateral (5% to 10%).[2,5] Posterolateral disc herniation may cause nerve root compression leading to back pain that progresses and radiates into the buttock, thigh, and leg in the distribution pattern of the involved nerve. A central disc herniation typically compresses the thecal sac while sparing the individual nerve root. This leads to low back pain due to sensory innervation to the meninges, posterior longitudinal ligament, and outer layers of the annulus fibrosus. Radiculopathy is usually absent unless the central disc herniation is so large that it compresses the cauda equina (Fig. 2C). A lateral disc herniation extends into or beyond the neural foramen with little or no impression on the dural sac. It may compress a nerve root as the nerve exits the neural foramen.

Almost all lumbar disc herniations occur at the lower three interspace levels[2,3] (Fig. 2D), and therefore a CT examination of L3-L4, L4-L5, and L5-S1 will almost always include the level of disc herniation. A patient with suspected lumbar disc herniation can be studied by CT with evaluation of the last three intervertebral discs (and intervening structures) unless the clinical data suggest that other levels be examined.

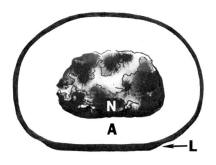

NORMAL

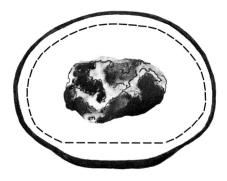

BULGING ANNULUS

SUBLIGAMENTOUS DISC HERNIATION

EXTRUDED DISC - ATTACHED

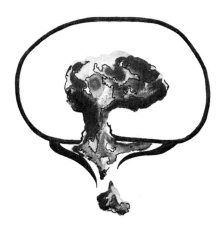

EXTRUDED DISC - FREE FRAGMENT

FIG. 2B. Disc disorders. A normal disc is shown at top. *N,* nucleus pulposus; *A,* annulus fibrosus; *L,* posterior longitudinal ligament. Bulging annulus: The annulus fibrosus protrudes in a generalized fashion beyond the vertebral body margin *(dotted line).* Subligamentous disc herniation: The nucleus pulposus ruptures through a tear in the annulus fibrosus but remains anterior to the posterior longitudinal ligament. Extruded disc: The nucleus pulposus ruptures through the annulus and extends posterior to the posterior longitudinal ligament. An extruded disc may remain attached to the parent disc or may be detached and lie within the spinal or neural canal as a free fragment. (Adapted from Burton in *Spine Update, 1984: Perspectives in Radiology, Orthopaedic Surgery, and Neurosurgery* (pp 107–112) by HK Genant (Ed), Radiology Research and Education Foundation, © 1983.)

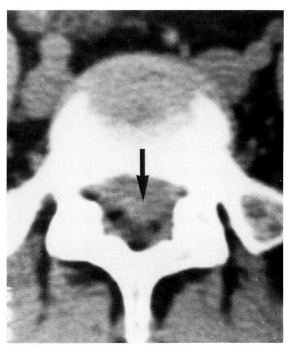
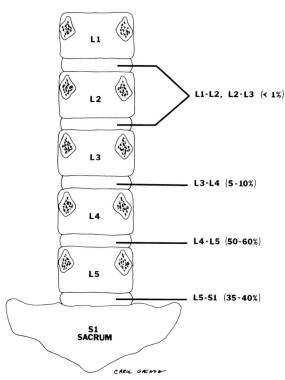

FIG. 2C. Large disc herniation. CT at L5-S1. This disc herniation *(arrow)* is predominantly central but is so large that there is extension to both sides of the midline. The thecal sac is compressed and displaced posteriorly.

FIG. 2D. Relative frequency of lumbar disc herniations. Data derived from Refs. 2 and 3.

References

1. Carrera GF, Williams AL, Haughton VM: Computed tomography in sciatica. *Radiology* 1980;137:433–437.
2. Fries JW, Abodeely DA, Vijungco JG, et al: Computed tomography of herniated and extruded nucleus pulposus. *J Comput Assist Tomogr* 1982;6:874–887.
3. Salenius P, Laurent LE: Results of operative treatment of lumbar disc herniation: A survey of 886 patients. *Acta Orthop Scand* 1977;48:630–634.
4. Williams AL, Haughton VM, Daniels DL, et al: Differential CT diagnosis of extruded nucleus pulposus. *Radiology* 1983;148:141–148.
5. Williams AL, Haughton VM, Syvertsen A: Computed tomography in the diagnosis of herniated nucleus pulposus. *Radiology* 1980;135:95–99.

CASE 3

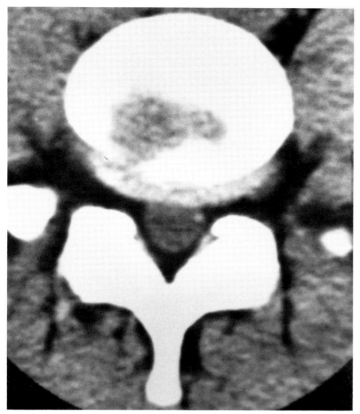

FIG. 3A. Axial CT at L5-S1.

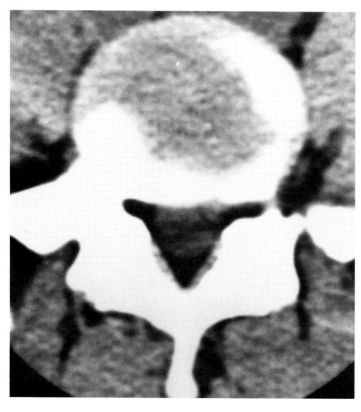

FIG. 3B. CT 4 mm caudad to Fig. 3A.

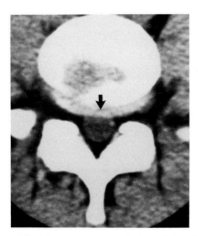 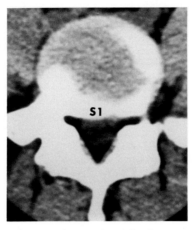

FIG. 3A-1. Pseudoherniation of the disc. The posterior margin of the disc *(arrow)* is convex and extends symmetrically posterior to the L5 vertebral body. The epidural fat is preserved.

FIG. 3B-1. The posterior aspect of S1 is demonstrated on the same scan as the mid and anterior portions of the L5-S1 disc. This is due to nonparallel scanning through the disc.

Pseudoherniation of the Disc

The posterior margin of the L5-S1 disc is slightly convex and projects beyond the margin of the L5 vertebral body (Fig. 3A-1). This simulates protrusion of the disc but is in fact a normal finding termed a pseudoherniation of the disc. It is produced by nonparallel scanning through the disc. Unlike in disc herniation, the pseudoherniation causes no compression or displacement of the thecal sac or nerve root. In addition, an evaluation of the scan obtained 4 mm caudad (Fig. 3B-1) reveals that the posterior aspect of the sacrum has the same extent and configuration as the posterior disc margin on the scan above (Fig. 3A-1).

In this case there are two findings that cause the psuedoherniation to simulate protrusion of the disc: The posterior margin of the disc is convex, and the disc extends posterior to the L5 vertebral body. First, it must be remembered that although the lumbar discs from L1-L2 to L4-L5 have flat or slightly concave posterior disc margins, it is normal for the L5-S1 disc to have a slightly convex posterior configuration. The second aspect of pseudoherniation of the disc concerns the appearance of the L5-S1 disc posterior to the L5 vertebral body. To understand the reason for this transaxial appearance let us examine the commonly employed methods of scanning.

There are two general methods of performing a lumbar CT scan.[1-3] Both methods begin with AP and lateral digital radiographs. The patient is in the supine position with the knees flexed. In one method the scans are obtained parallel to the disc space by appropriate angulation of the gantry. The examination includes scan slices that are 3, 4 or 5 mm thick. These are obtained through and about the disc and include at least part of the surrounding pedicles. Approximately seven scans are obtained at each level. The L3-L4, L4-L5, and L5-S1 discs are typically studied in this fashion with individual gantry angulation (Fig. 3C). In the second general method of scanning, no attempt is made to scan parallel to the disc. Instead, contiguous scanning is performed from approximately the level of the L3 pedicle to the superior aspect of the sacrum using a 0° angulation of the gantry (Fig. 3D). Some investigators use contiguous scanning with additional angled scanning at the L5-S1 level.

The maximum angle that the CT gantry can assume is 15° to 25°, depending on the equipment. Although the maximum gantry angle is almost always sufficient to examine parallel to the first four lumbar disc spaces, the angle of the L5-S1 disc often exceeds the maximum gantry angle, thus preventing parallel scanning through the L5-S1 disc. Some authors have recommended scanning while the patient is prone. This diminishes the lumbar lordosis and increases the number of patients in whom parallel scanning can be performed at L5-S1.[5] However, this method is not widely practiced, primarily because of patient discomfort, which can lead to difficulty in maintaining proper positioning. When the angle of the intervertebral disc exceeds the gantry angle (either because a 0° angle is chosen or because the disc space angle

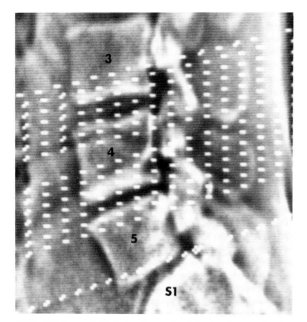 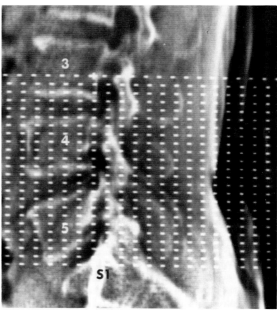

FIG. 3C. Lateral digital radiograph of the lumbosacral spine demonstrating the method of scanning parallel to the disc. Individual gantry angulation was used to obtain scans parallel to the L3-L4 and L4-L5 intervertebral discs. The single cursor line at L5-S1 denotes the angulation of scans that are to be obtained subsequently above, below, and through the L5-S1 disc. Note that the maximum 20° gantry angulation was used at L5-S1 but is insufficient for parallel scanning through the L5-S1 disc.

FIG. 3D. Lateral digital radiograph of the lumbosacral spine with cursor lines demonstrating levels of scans obtained in the contiguous method of scanning. With this method, scans are obtained from the midportion of L3 to the superior aspect of the sacrum at regular intervals, with no attempt to scan parallel to the disc.

exceeds the maximum gantry angle), the disc projects posterior to the superior vertebral body. This occurs frequently at L5-S1 with the intervertebral disc visualized posterior to L5. A scan obtained just caudad to this level demonstrates the midportion of the L5-S1 disc with the L5 vertebral body anterior and the sacrum posterior (Figs. 3E–G). A similar but more dramatic CT appearance of a pseudoherniation of the disc occurs when spondylolisthesis is present. Scoliosis is another cause of pseudoherniation. The left and right sides of the intervertebral disc are scanned asymmetrically in patients with scoliosis. This leads to the impression that disc material extends focally beyond the vertebral body margin, thus suggesting a posterolateral or lateral disc herniation. Evaluation of the AP digital radiograph confirms the presence of scoliosis. In this setting caution must be exercised in diagnosing disc herniation.

Now that we have discussed the pseudoherniation of the disc, a pitfall of CT evaluation of herniation of the nucleus pulposus (HNP), let us come back to the question of scanning technique. What is the best method of performing the lumbar CT study? There is no convincing consensus of opinion, and clearly both methods are acceptable. Like others,[1] we prefer to scan parallel to the disc space when the patient's history is suggestive of HNP, especially in younger patients in whom osseous disease is less likely to be present. We use the contiguous scanning method with 0° angulation for the evaluation of osseous stenosis, spondylolisthesis, spondylolysis, trauma, and tumor. The contiguous scanning method prevents the possibility of inadvertent failure to scan through a pathologic process at the midvertebral level. This non-angled technique is also useful in obtaining reconstructed images in the sagittal and coronal planes. Some computer programs are capable of reformatting a reformatted sagittal image into a plane exactly parallel to the intervertebral disc.[4] The use of an overlapping technique (e.g., 5-mm slice thickness with scans obtained at 3- or 4-mm intervals) helps achieve better quality reconstruction views. The overlapping method, however, increases the number of scans obtained and therefore increases radiation dosage and examination time. These factors must be weighed against the need to obtain optimal reconstruction views when the examiner determines the technique to be used for the individual patient.

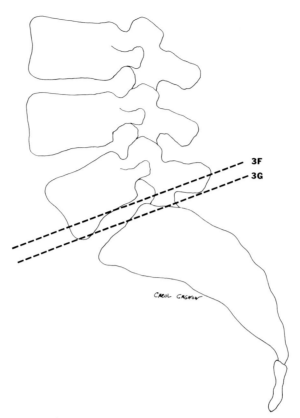

FIG. 3E. Lateral view of the lumbar spine. *Dotted lines* indicate the planes of scanning represented in Figs. 3F and 3G. Note that these lines are not parallel to the intervertebral disc despite maximum 20° angulation of the gantry.

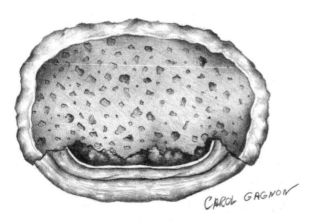

FIG. 3F. Structures scanned through the plane of line 3F in Fig. 3E. The L5-S1 disc projects posterior to the L5 vertebral body because the angle of the intervertebral disc space exceeds the maximum angle of the CT gantry.

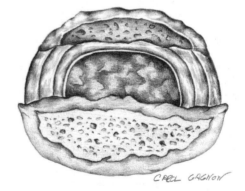

FIG. 3G. Structures scanned through the plane of line 3G in Fig. 3E. The L5-S1 disc is surrounded by the inferior aspect of L5 anteriorly and the superior aspect of S1 posteriorly.

References

1. Braun IF, Lin JP, George AE, et al: Pitfalls in the computed tomographic evaluation of the lumbar spine in disc disease. *Neuroradiology* 1984;26:15–20.
2. Genant HK: Computed tomography of the lumbar spine: Technical considerations, in Genant HK, Chafetz N, Helms CA (eds): *Computed Tomography of the Lumbar Spine.* San Francisco, University of California, 1982, pp 23–52.
3. Haughton VM, Williams AL: *Computed Tomography of the Spine.* St Louis, CV Mosby, 1982.
4. Hirschy JC, Leue WM, Berninger WH, et al: CT of the lumbosacral spine: Importance of tomographic planes parallel to vertebral end plate. *AJNR* 1980;1:551–556; *AJR* 1981;136:47–52.
5. Tehranzadeh J, Gabriele OF: The prone position for CT of the lumbar spine. *Radiology* 1984;152:817–818.

CASE 4

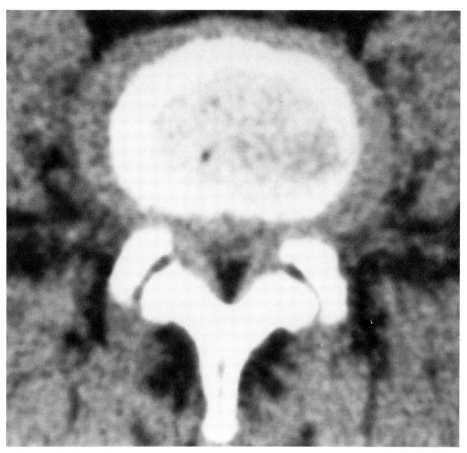

FIG. 4A. Axial CT at L3-L4 in a 71-year-old male with chronic back pain.

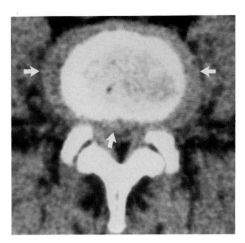

FIG. 4A-1. Bulging annulus. There is generalized bulging of the annulus fibrosus *(arrows)* beyond the vertebral body margins. No focal herniation is identified. Some compression of the thecal sac is present.

Bulging Annulus

There is bulging of the annulus fibrosus, which protrudes in a generalized fashion beyond the vertebral body (Fig. 4A-1). CT can be used to accurately diagnose bulging of the annulus fibrosus and differentiate it from herniation of the nucleus pulposus.[1,4,5] CT demonstrates a generalized, usually symmetric extension of the disc margin beyond the vertebral body, with disc protruding posteriorly, laterally, and anteriorly (Fig. 4B). There is no focal protrusion of the disc margin as occurs with herniation of the nucleus pulposus.[4,5] The posterior margin of a bulging disc is most often convex but infrequently may remain concave (Fig. 4C) because of reinforcement of the central portion of the annulus by the posterior longitudinal ligament.[4,5] Gas within the disc and calcification of the outer annulus may be present.

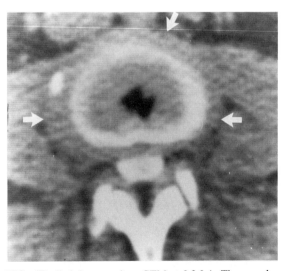

FIG. 4B. Bulging annulus. CTM at L3-L4. The annulus bulges in a generalized fashion anteriorly, posteriorly, and laterally *(arrows)*. The anterior aspect of the thecal sac is flattened by the bulging annulus. There is calcification within the periphery of the annulus on the right. Note low-density gas in the disc (vacuum disc phenomenon) indicating degeneration of the disc.

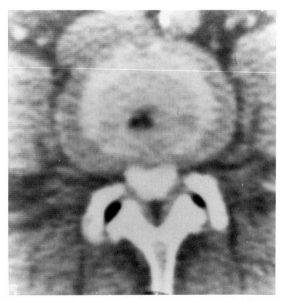

FIG. 4C. CTM at L4-L5 in the same patient as in Fig. 4B. At this level the bulging of the annulus is not as prominent in the central posterior portion as it is elsewhere. This is because of reinforcement of the central portion of the annulus by the posterior longitudinal ligament. Note gas in the facet joints (vacuum facet) and in the disc (vacuum disc phenomenon).

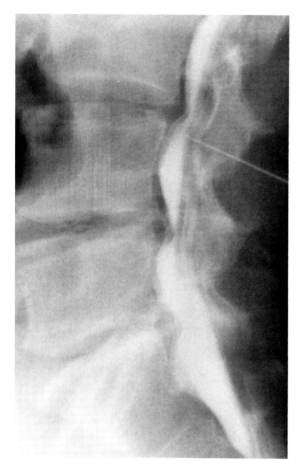

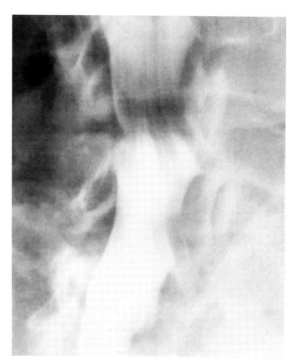

FIG. 4E. Bulging of the annulus fibrosus. Oblique view from a myelogram of the same patient as in Fig. 4D. There is uniform thinning of the contrast within the thecal sac at the level of the L4-L5 intervertebral disc. The nerve root sleeves fill with contrast and are not amputated or widened as might occur with disc herniation.

FIG. 4D. Bulging of the annulus fibrosus at multiple levels. Cross-table lateral view from a myelogram performed with water-soluble contrast. Anterior rounded extradural defects are causing compression of the thecal sac at L3-L4, L4-L5, and L5-S1. The needle is at the L3-L4 level.

Bulging of the annulus fibrosus can also be accurately diagnosed when myelography is performed with water-soluble contrast. The bulging disc appears as a rounded, usually symmetric, bilateral extradural deformity of the thecal sac without extension caudad or cephalad from the level of the disc[2] (Figs. 4D, 4E). In addition, the nerve roots are uniform in caliber and normal in size. Compare these myelographic findings of generalized bulging of the disc to the abnormalities found on the CT study of the same patient (Fig. 4F).

Pathologically, several biochemical and biomechanical factors may lead to generalized bulging of the disc with aging.[4,5] Gradual desiccation of the nucleus pulposus leads to decreased nuclear turgor, permitting a decrease in disc space height. In addition, the annulus fibrosus develops fissuring, hyalin degeneration, and increased pigmentation. The annulus loses elasticity and bulges in a generalized fash-

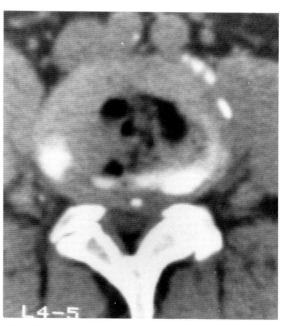

FIG. 4F. Bulging annulus at L4-L5. The annulus bulges in a generalized fashion, compressing the thecal sac. This finding corresponds to that demonstrated myelographically in Figs. 4D and 4E from the same case. There is calcification in the periphery of the annulus. A vacuum disc phenomenon is noted.

ion beyond the adjacent vertebral body margins.[4,5] The inner fibers of the annulus fibrosus may tear; however, the outer fibers remain intact, preventing any nuclear material from herniating.[5]

Clinically, patients with generalized bulging of the disc are unlikely to have nerve root compression. However, in some instances a bulging disc may be associated with thickening of the ligamenta flava, prominent laminae, and hypertrophied articular processes.[2] This combination of abnormalities may lead to spinal stenosis with nerve root compression that is predominantly related to bony and ligamentous encroachment.[3]

References

1. Haughton VM, Eldevik OP, Magnaes B, et al: A prospective comparison of computed tomography and myelography in the diagnosis of herniated lumbar disks. *Radiology* 1982;142:103–110.
2. Kieffer SA, Sherry RG, Wellenstein DE, et al: Bulging lumbar intervertebral disk: Myelographic differentiation from herniated disk with nerve root compression. *AJR* 1982;138:709–716.
3. Roberson GH, Llewellyn HJ, Taveras JM: The narrow lumbar spinal canal syndrome. *Radiology* 1973;107:87–97.
4. Williams AL: CT diagnosis of degenerative disc disease: The bulging annulus. *Radiol Clin North Am* 1983;21:289–300.
5. Williams AL, Haughton VM, Meyer GA, et al: Computed tomographic appearance of the bulging annulus. *Radiology* 1982;142:403–408.

CASE 5

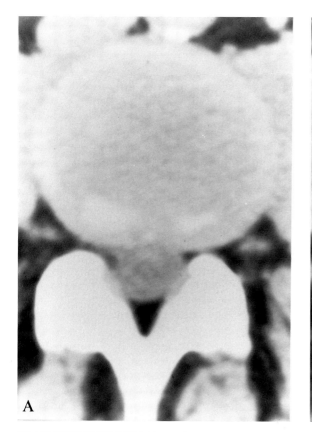

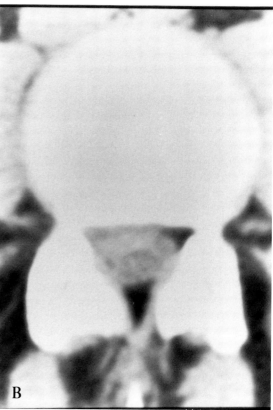

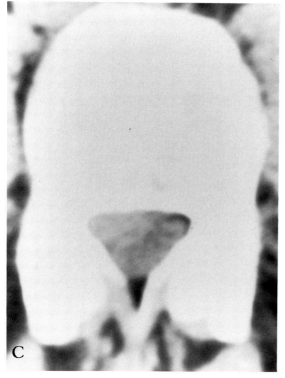

FIG. 5A. Axial CT parallel to and through the L3-L4 intervertebral disc. This 56-year-old male has back pain and a right radiculopathy.

FIG. 5B. CT 5 mm caudad to Fig. 6A.

FIG. 5C. CT 10 mm caudad to Fig. 6A.

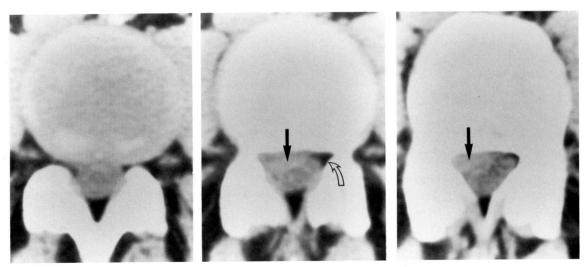

FIG. 5A-1. Normal-appearing disc. There is no evidence of disc herniation on the scan obtained parallel to and through the intervertebral disc.

FIG. 5B-1. Extruded disc. CT 5 mm caudad to the disc demonstrates an extruded disc on the right *(straight arrow)* compressing the thecal sac and causing obliteration of the epidural fat. Compare with normal epidural fat on the left *(curved arrow)*.

FIG. 5C-1. Extruded disc *(arrow)* is still visualized 10 mm caudad to the midportion of the intervertebral disc.

Extruded Disc

The posterior margin of the disc appears normal on the transaxial scan obtained parallel to and through the disc (Fig. 5A-1). However, posterolateral disc herniation is seen on scans obtained 5 and 10 mm caudad to the center of the disc space (Figs. 5B-1, 5C-1). At surgery, an extruded disc was found.

An extruded disc is present when nuclear material protrudes through a tear in the annulus fibrosus and extends through or around the posterior longitudinal ligament.[2,7] The portion of the disc that has extruded beyond the ligament may remain solidly attached to the parent disc or may be displaced and lie as a free fragment within the vertebral canal or neural foramen. In one large series of patients with disc herniation, 35% had an extruded disc at surgery.[5]

Attempting to distinguish extruded disc from subligamentous disc herniation by CT is not always successful; however, Table 5-1 lists the differential features. An extruded disc can be diagnosed when a free fragment is seen displaced into the canal and separated from the disc margin by epidural fat (Fig. 5D), a finding that occurs in only 50% of the cases.[4,7] In CT examination of the other half of the cases, the extruded disc is contiguous with the posterior margin and may be indistinguishable from a focal subligamentous herniation, particularly when the contour is smooth and curvilinear. An extruded disc may be considered if the soft-tissue mass has a polypoid or irregular shape. A free disc fragment may migrate superiorly (Fig. 5E), inferiorly (Fig. 5F), or in both directions and may be as far as 15 to 30 mm from the intervertebral disc space. An extruded fragment has been reported to lie 6 mm or more from the center of the disc space in 85% of cases of disc extrusion[4] (Fig. 5G). The fragment may appear larger in the cephalad or caudad direction than at the disc level—

TABLE 5-1 Differential Features of Extruded Disc and Subligamentous Disc Herniation

Feature	Extruded Disc	Subligamentous Disc Herniation
Separation from posterior disc	50% are separated from disc margin by fat[4,7]	Contiguous with remainder of disc
Cephalad or caudad displacement	85% >6 mm from center of disc[4]	Most prominent at disc space level
Size	Often large[5]	Usually not large[5]
Shape	Irregular, polypoid	Smooth, curvilinear

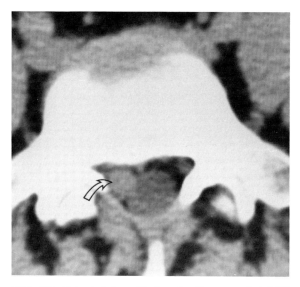

FIG. 5D. Extruded disc with free disc fragment. Axial CT obtained 8 mm caudad to the L5-S1 intervertebral disc space. There is a large free disc fragment *(arrow)* displaced caudad, which is causing compression of the thecal sac on the right. The free fragment is also displaced posteriorly into the spinal canal and is separated from the posterior aspect of the vertebral body by a thin layer of epidural fat.

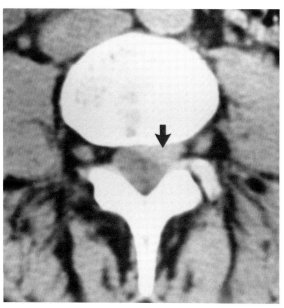

FIG. 5E. Extruded disc. Axial CT 8 mm cephalad to the midportion of the L4-L5 intervertebral disc at the level of the neural foramina. The extruded disc *(arrow)* is displaced cephalad and is causing compression of the thecal sac and obliteration of the epidural fat.

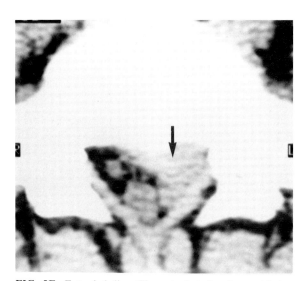

FIG. 5F. Extruded disc. The extruded disc *(arrow)* is located 10 mm caudad to the midportion of the L5-S1 intervertebral disc. There is posterior displacement of the left S1 nerve root and obliteration of the anterior epidural fat. An extruded disc rather than a subligamentous herniation is suggested by the large size and considerable caudad displacement of the disc material. This was confirmed at surgery.

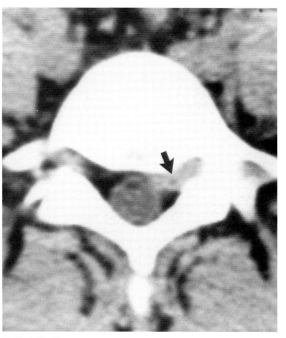

FIG. 5G. Extruded disc with free disc fragment. The extruded disc *(arrow)* is displaced 12 mm cephalad to the midportion of the L5-S1 intervertebral disc. It extends laterally into the left neural foramen, compressing the L5 nerve root. Surgical exploration was guided by the CT findings, and a large free disc fragment was found and removed from the superior aspect of the neural foramen.

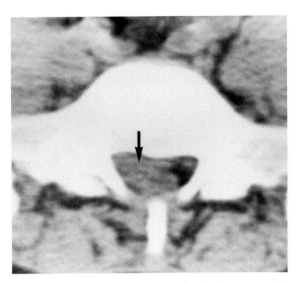

FIG. 5H. Extruded disc. This extruded disc *(arrow)* is displaced 12 mm caudad to the midportion of the L5-S1 intervertebral disc. The AP diameter of the disc material is greater than one half the anticipated normal sagittal diameter of the dural sac. The large size and marked caudad displacement suggest that the disc has extruded beyond the posterior longitudinal ligament. This was confirmed at surgery.

a finding not present in patients with subligamentous herniation.[4] It is important to study scans above and below the disc space, since approximately 10% of disc extrusions are associated with a normal posterior disc margin and would go undetected if evaluation were limited to the disc itself[4,7] (Fig. 5A-1). Extruded discs tend to be larger than subligamentous herniations. In one study the maximum AP diameter of the herniated disc was compared to the anticipated normal sagittal diameter of the dural sac. In patients in whom the ratio was less than one half, 10% had an extruded disc at surgery. In patients with a ratio of one half or more, 90% had an extruded disc[5] (Fig. 5H).

The CT diagnosis of a free fragment is significant. It alerts the surgeon to the presence and location of the fragmented disc, and it suggests a less favorable outcome should chymopapain chemonucleolysis be attempted.[1,6] Rarely, a free fragment tears the dura and arachnoid and enters the subarachnoid space as an intradural disc herniation and causes compression of the cauda equina.[3,4]

References

1. Benoist M, Deburge A, Busson J, et al: Treatment of lumbar disc herniation by chymopapain chemonucleolysis. *Spine* 1982;7:613–617.
2. Carrera GF, Williams AL, Haughton VM: Computed tomography in sciatica. *Radiology* 1980;137:433–437.
3. Ciappetta P, Delfini R, Cantore GP: Intradural lumbar disc hernia: Description of three cases. *Neurosurgery* 1981;8:104–107.
4. Dillon WP, Kaseff LG, Knackstedt VE, et al: Computed tomography and differential diagnosis of the extruded lumbar disc. *J Comput Assist Tomogr* 1983;7:969–975.
5. Fries JW, Abodeely DA, Vijungco JG, et al: Computed tomography of herniated and extruded nucleus pulposus. *J Comput Assist Tomogr* 1982;6:874–887.
6. Gentry LR, Strother CM, Turski PA, et al: Chymopapain chemonucleolysis: Correlation of diagnostic radiographic factors and clinical outcome. *AJR* 1985;145:351–360.
7. Williams AL, Haughton VM, Daniels DL, et al: Differential CT diagnosis of extruded nucleus pulposus. *Radiology* 1983;148:141–148.

CASE 6

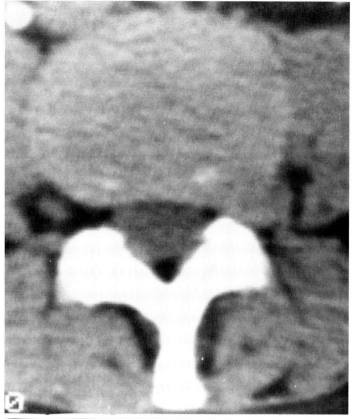

FIG. 6A. Axial CT scan through the L3-L4 intervertebral disc. The patient had a 4-week history of left lower back pain and left radiculopathy.

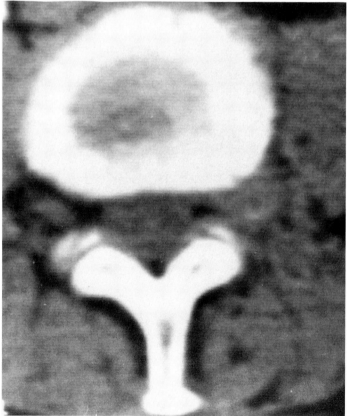

FIG. 6B. CT 4 mm above the L3-L4 disc.

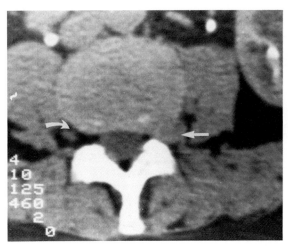
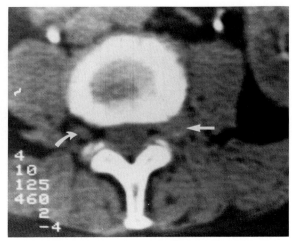

FIG. 6A-1. A focal lateral disc herniation *(straight arrow)* causes obliteration of the perineural fat. Compare with normal perineural fat on the right *(curved arrow).* There is no compression of the thecal sac.

FIG. 6B-1. Above the disc space at the level of the inferior portion of the neural foramen. The lateral disc herniation *(straight arrow)* has extended cephalad. The left L3 nerve root cannot be visualized because of the large disc herniation. Compare with the normal nerve root on the right *(curved arrow).*

Lateral Disc Herniation

CT makes an important contribution to the diagnosis of disc disease with its ability to depict far lateral disc herniations that might otherwise go undetected by myelography or limited surgical exploration.[3–5,8,12] In this case there is focal protrusion of the disc margin at L3-L4 far lateral on the left causing obliteration of perineural fat within the neural foramen (Figs. 6A-1, 6B-1). The L3 nerve root exits through the upper half of the neural foramen located beneath the L3 pedicle. A lateral disc herniation at L3-L4 may compress the exiting L3 nerve root if the disc is displaced cephalad.

The most frequent CT findings associated with lateral disc herniation include focal protrusion of disc within or lateral to the neural foramen, displacement of fat within the foramen, and absence of dural sac deformity.[12] Less frequently, disc herniation may protrude to a greater extent and appears as a large soft-tissue mass within or lateral to the foramen. The density of herniated disc material is about the same or slightly less than that of the intervertebral disc and almost always greater than that of the thecal sac.[3,9] Occasionally, calcification or gas may be seen within the herniated lateral disc. Approximately 5% of all lumbar disc herniations diagnosed by CT are far lateral.[2,8,12]

Lateral disc herniation is a known cause of false-negative myelograms.[4,10,11] Unlike CT with its direct anatomic visualization, myelography relies on indirect evidence of disease by uncovering the effects on the thecal sac and nerve roots. With metrizamide myelography, the nerve root sheaths can be visualized to their termination near the dorsal root ganglion within the neural foramen. Disc herniation located lateral to the ganglion may therefore go undetected with myelography. An example of a far lateral disc herniation at L4-L5 is shown in a patient who presented with right radiculopathy and had a normal myelographic examination performed with water-soluble contrast (Fig. 6C).

Information gained by CT may be of great benefit in the surgical management of patients with lateral disc herniation. The specific compromised nerve root can often be suggested by the pain distribution, physical examination, and electromyography.[1] However, it must be kept in mind that a far lateral disc herniation compromises the same nerve root as does a typical posterolateral disc herniation at the next higher level (Table 6-1 and Fig. 6D). When the far lateral position of a disc herniation is not appreciated preoperatively, initial surgical exploration may be unrewarding[7] (Figs. 6E, 6F).

The differential diagnosis of lateral disc herniation includes schwannoma and soft-tissue involvement by lymphoma, metastatic disease, or infection.[3,9,12] Cystic nerve root sleeve dilatation and conjoined nerve root sheath anomaly may also be considered. A schwannoma located in the interver-

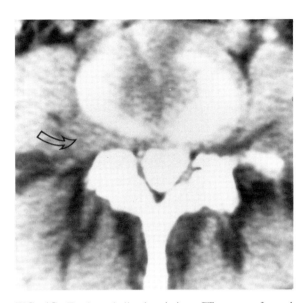

FIG. 6C. Far lateral disc herniation. CT scan performed several hours after myelography with water-soluble contrast. There is a large far lateral disc herniation *(arrow)* at L4-L5. Despite the size of the herniation, a myelogram with water-soluble contrast was unremarkable because the herniation was lateral to the myelographically visualized portion of the nerve root sheath.

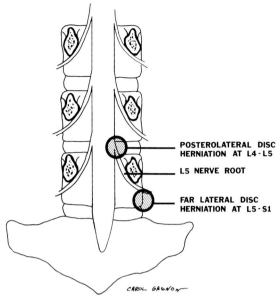

FIG. 6D. Effect of posterolateral disc herniation at L4-L5 and far lateral disc herniation at L5-S1 on the same L5 nerve root.

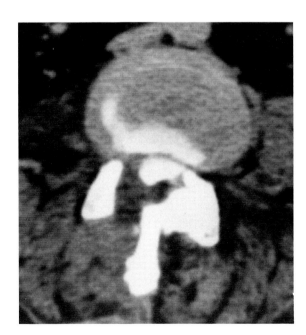

FIG. 6E. Postoperative CTM. This patient remained symptomatic after surgery. Scan is at the L3-L4 intervertebral disc level. Note right laminectomy.

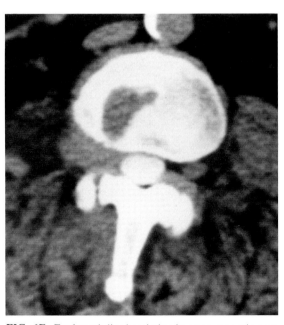

FIG. 6F. Far lateral disc herniation in a symptomatic postoperative patient (same patient as in Fig. 6E). Cephalad to disc level a large lateral disc herniation is seen compressing the dorsal root ganglion of L3 on right. The far lateral extent of the disc herniation had not been appreciated prior to initial surgery. Information gained from CT guided reoperation.

Case 6

TABLE 6-1 Effect of Posterolateral and Far Lateral Herniation of the Nucleus Pulposus on the Lumbar Nerve Root

Nerve Root	Posterolateral HNP	Far Lateral HNP	Diminished or Absent Reflex[1]	Pain and Paresthesias[1]
L3	L2-L3	L3-L4	Knee	Anterior thigh and knee
L4	L3-L4	L4-L5	Knee	Anterior thigh and knee, medial leg
L5	L4-L5	L5-S1	± Ankle	Hip, posterolateral thigh, lateral calf, dorsal foot, 1st or 2nd and 3rd toes
S1	L5-S1		Ankle	Midgluteal, posterior thigh, posterior calf to heel, outer plantar foot, 4th and 5th toes

tebral foramen may have a sharp border similar to that seen in lateral disc herniation. However, a schwannoma may demonstrate enhancement after intravenous injection of iodinated contrast[3] and may cause widening of the neural foramen. Metastasis and lymphoma usually have attenuation values similar to or slightly less than those of a herniated lateral disc. However, these neoplasms are likely to have irregular, indistinct margins with an infiltrative appearance and may cause widening of paraspinal soft tissues.[3,9] Metastasis is usually associated with bone destruction of the adjacent pedicle or vertebral body. Similarly, soft-tissue infection has indistinct margins and may be associated with adjacent bone destruction. Cystic dilatation of the nerve root sleeve is isodense with the thecal sac because it contains cerebrospinal fluid. Unlike in lateral disc herniation, the density of the conjoined nerve root sheath is similar to or only slightly greater than that of the thecal sac. Serial scans may show the derivation of the conjoined nerve roots from the dural sac and the asymmetry of the nerve roots.[6]

References

1. Adams RD, Victor M: *Principles of Neurology*, ed 3. New York, McGraw-Hill, 1985.
2. Fries JW, Abodeely DA, Vijungco JG, et al: Computed tomography of herniated and extruded nucleus pulposus. *J Comput Assist Tomogr* 1982;6:874–887.
3. Gado M, Patel J, Hodges FJ III: Lateral disk herniation into the lumbar intervertebral foramen: Differential diagnosis. *AJNR* 1983;4:598–600.
4. Godersky JC, Erickson DL, Seljeskog EL: Extreme lateral disc herniation: Diagnosis by computed tomographic scanning. *Neurosurgery* 1984;14:549–552.
5. Haughton VM, Eldevik OP, Magnaes B, et al: A prospective comparison of computed tomography and myelography in the diagnosis of herniated lumbar disks. *Radiology* 1982;142:103–110.
6. Helms CA, Dorwart RH, Gray MG: The CT appearance of conjoined nerve roots and differentiation from a herniated nucleus pulposus. *Radiology* 1982;144:803–807.
7. Macnab I: Negative disc exploration: An analysis of the causes of nerve-root involvement in sixty-eight patients. *J Bone Joint Surg Am* 1971;53-A:891–903.
8. Novetsky GJ, Berlin L, Epstein AJ, et al: The extraforaminal herniated disk: Detection by computed tomography. *AJNR* 1982;3:653–655.
9. Schubiger O, Valavanis A, Hollmann J: Computed tomography of the intervertebral foramen. *Neuroradiology* 1984;26:439–444.
10. Shapiro R. *Myelography*, ed 4. Chicago, Year Book Medical Publishers, 1984.
11. Strother CM: Lumbar examination, in Sackett JF, Strother CM (eds): *New Techniques in Myelography*. Philadelphia, Harper and Row, 1979, pp 69–89.
12. Williams AL, Haughton VM, Daniels DL, et al: CT recognition of lateral lumbar disk herniation. *AJNR* 1982;3:211–213, *AJR* 1982;139:345–347.

CASE 7

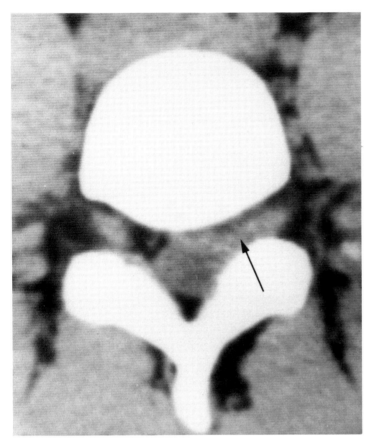

FIG. 7A. Axial CT scan 6 mm cephalad to the L5-S1 intervertebral disc. Density measurement at site of *arrow* is 28 HU. This 27-year-old male had left-sided low back and leg pain.

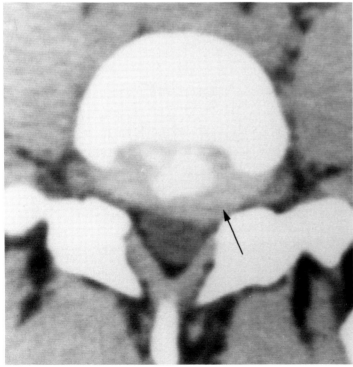

FIG. 7B. CT scan at the L5-S1 intervertebral disc. Density measurement at site of *arrow* is 59 HU.

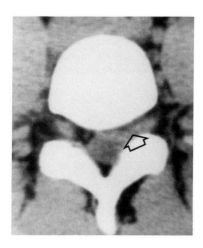
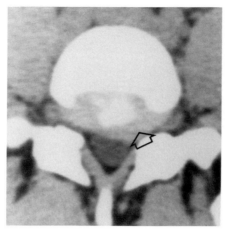

FIG. 7A-1. Conjoined nerve roots. A soft-tissue density of conjoined L5 and S1 nerve roots *(arrow)* is seen within the left neural foramen. The conjoined nerve roots measured 28 HU. There is asymmetry of the epidural fat.

FIG. 7B-1. Disc herniation. At the level of the intervertebral disc, 6 mm caudad to Fig. 7A-1, there is a left posterolateral disc herniation *(arrow)*, which measured 59 HU.

Conjoined Nerve Roots and Disc Herniation

The first scan is at the level of the neural foramina (Fig. 7A-1). A soft-tissue density is seen extending into the left foramen, causing obliteration of the perineural fat. Six millimeters caudad, at the level of the disc, another soft-tissue density is seen extending into the anterolateral aspect of the spinal canal on the left side (Fig. 7B-1). The two soft-tissue structures have different CT density measurements and represent two distinct entities. High-resolution CT scanners are equipped with the ability to "highlight" or "blink in" those tissues having density measurements that fall within a preselected range of Hounsfield units. When Fig. 7A-1 is examined with highlighting in the range of 0 to 30 HU, the soft-tissue density within the foramen blinks in along with the thecal sac (Fig. 7C). This indicates that the soft tissue within the foramen is similar in density to the thecal sac and nerve roots and in fact represents conjoined nerve roots. At the disc level (Fig. 7B-1) the soft tissue extending into the canal has density measurements similar to those of disc and blinks in when a range of 50 to 100 HU is selected (Fig. 7D). This is due to disc herniation. Although myelography demonstrated conjoined nerve roots without evidence of disc herniation, at surgery the diagnosis of both conjoined nerve roots and disc herniation was confirmed.

Conjoined nerve roots are congenital anomalies in which two nerve roots emerge from a common dural sheath. The myelographic appearance of conjoined nerve roots is shown in Fig. 7E. Conjoined nerve roots have been reported in 1% of lumbar disc operations[1,8] and 8% of anatomic specimens.[5] The nerve roots that are most frequently conjoined are L5 and S1. Less often the L4 and L5 roots or the S1 and S2 roots are involved. The nerve roots may exit from the same foramen, or they may exit separately with the bifurcation of the conjoined nerve roots in close approximation to the intervening pedicle[8] (Figs. 7F, 7G). Conjoined nerve roots occurring without coexistent disc herniation are usually not the cause of symptoms,[3] although sciatica has been reported when neural canal stenosis is present. However, a herniated disc and conjoined nerve roots may occur together as in the present case. When a disc herniation is located at the site of conjoined nerve roots, symptoms may be present at more than one root level.

With CT, conjoined nerve roots appear as a soft-tissue density in the anterior epidural space, neural foramen, and lateral recess, causing obliteration of the epidural and perineural fat (Fig. 7H). This may simulate a herniation of the nucleus pulposus (HNP) with lateral extension, but careful evaluation will lead to an accurate diagnosis.[3,7] Hounsfield density measurements are important since conjoined nerve roots are similar to or only slightly greater in density than the thecal sac, whereas the density of HNP is signif-

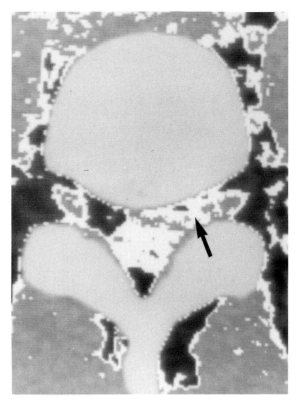

FIG. 7C. Conjoined nerve roots. Same CT scan as in Fig. 7A-1 studied with highlighting of tissues with density measurements of 0 to 30 HU. Note that the conjoined nerve roots *(arrow)* and the thecal sac are highlighted.

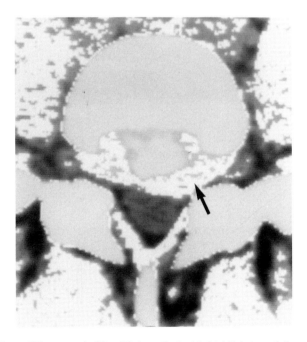

FIG. 7D. Disc herniation. Same CT scan as in Fig. 7B-1 studied with highlighting of tissues with density measurements of 50 to 100 HU. The left posterolateral disc herniation *(arrow)* is highlighted along with other structures of similar density measurements such as the ligamentum flavum and the paravertebral muscles. Note that the thecal sac and nerve roots are not highlighted because they are within a lower density range as shown in Fig. 7C.

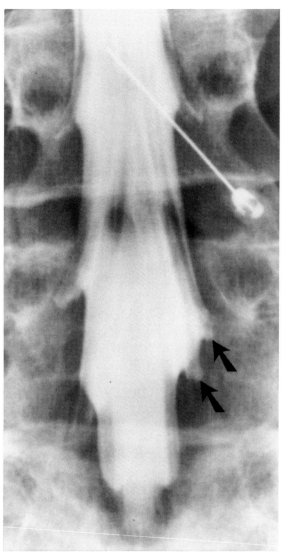

FIG. 7E. Conjoined nerve roots. Myelography with water-soluble contrast demonstrates conjoined L5 and S1 nerve roots *(arrows)*.

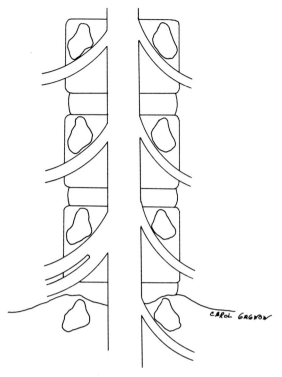

FIG. 7F. Conjoined nerve roots. The nerve roots share a common dural sheath and then exit together beneath the pedicle.

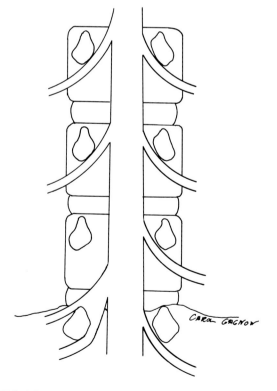

FIG. 7G. Conjoined nerve roots exiting separately in close apposition to the intervening pedicle.

icantly higher, similar to disc measurements. In general, the thecal sac and nerve roots have a range of 0 to 30 HU whereas disc densities range from approximately 50 to 100 HU. The blink mode can be used to give a pictorial representation of the density measurements (Fig. 7I). Careful comparison of the nerve roots on each side will detect asymmetry and may aid in the diagnosis of conjoined nerve roots. Slight dilatation of the ipsilateral lateral recess has been demonstrated by CT in patients with conjoined nerve roots.[4] This asymmetry of the bony spinal canal may be another diagnostic clue to the presence of conjoined nerve roots.

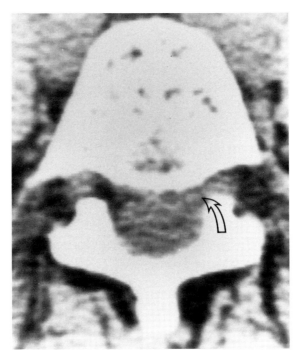

FIG. 7H. Conjoined nerve roots. Axial CT scan 8 mm above the L5-S1 disc. Conjoined L5 and S1 nerve roots *(arrow)* within the left neural foramen obliterate epidural fat. Same patient as in Fig. 7E.

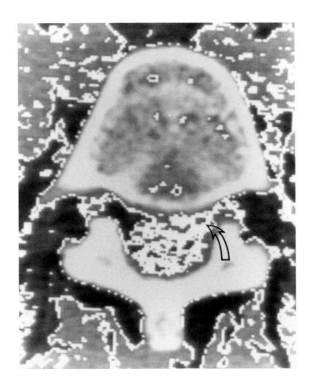

FIG. 7I. Conjoined nerve roots. Same CT scan as in Fig. 7H studied by highlighting of tissues measuring 0 to 30 HU. The conjoined nerve roots *(arrow)* are highlighted along with the thecal sac and the opposite nerve root, which have similar measurements.

It is important to be familiar with the CT appearance of conjoined nerve roots so as to avoid a mistaken diagnosis of disc herniation. When disc herniation and conjoined nerve roots coexist, the preoperative recognition of conjoined nerve roots may help avoid inadvertent nerve root damage at the time of disc surgery. In this clinical setting, decompressive foraminotomy or pediculectomy has been recommended in conjunction with discectomy and hemilaminectomy.[6,8]

In addition to disc herniation, the differential diagnosis of conjoined nerve roots on CT examination includes neural sheath tumor, metastasis that is isodense with the thecal sac, and postoperative fibrosis.[3] If the diagnosis of conjoined nerve roots is in doubt, myelography with water-soluble contrast is diagnostic. In some cases it is necessary to correlate both myelographic and CT findings to obtain a definitive diagnosis of coexistent HNP and conjoined nerve roots.[2]

References

1. Coughlin JR, Miller JDR: Metrizamide myelography in conjoined lumbosacral nerve roots. *J Can Assoc Radiol* 1983;34:23–25.
2. Gebarski SS, McGillicuddy JE: "Conjoined" nerve roots: A requirement for computed tomographic and myelographic correlation for diagnosis. *Neurosurgery* 1984;14:66–68.
3. Helms CA, Dorwart RH, Gray MG: The CT appearance of conjoined nerve roots and differentiation from a herniated nucleus pulposus. *Radiology* 1982;144:803–807.
4. Hoddick WK, Helms CA: Bony spinal canal changes that differentiate conjoined nerve roots from herniated nucleus pulposus. *Radiology* 1985;154:119–120.
5. Kadish LJ, Simmons EH: Anomalies of the lumbosacral nerve roots: An anatomical investigation and myelographic study. *J Bone Joint Surg Br* 1984;66-B:411–416.
6. Neidre A, Macnab I: Anomalies of the lumbosacral nerve roots: Review of 16 cases and classification. *Spine* 1983;8:294–299.
7. Peyster RG, Teplick JG, Haskin ME: Computed tomography of lumbosacral conjoined nerve root anomalies: Potential cause of false-positive reading for herniated nucleus pulposus. *Spine* 1985;10:331–337.
8. White JG, Strait TA, Binkley JR, et al: Surgical treatment of 63 cases of conjoined nerve roots. *J Neurosurg* 1982;56:114–117.

CASE 8

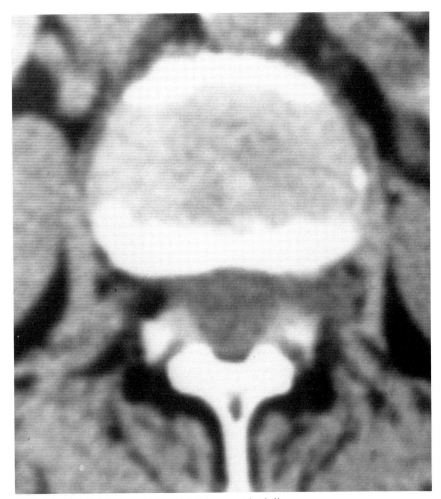

FIG. 8A. Axial CT through the L1-L2 intervertebral disc.

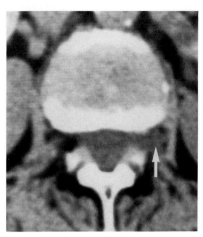

FIG. 8A-1. Cystic dilatation of the nerve root sleeve *(arrow)* within the left neural foramen appears rounded and isodense with the thecal sac. This measured 14 HU. Note that the perineural fat is obliterated.

Cystic Nerve Root Sleeve Dilatation

There is a large, round "mass" in the left neural foramen causing obliteration of the perineural fat (Fig. 8A-1). The CT density of this abnormality appears visually to be similar to that of the thecal sac and much less than that of the disc. Measurements confirm that both the thecal sac and the round "mass" have a similar density of 14 HU. This is due to cystic nerve root sleeve dilatation (CNRSD). A metrizamide myelogram and CTM performed for other clinical reasons confirmed the presence of CNRSD at multiple lumbar levels (Figs. 8B, 8C).

As the normal nerve root emerges from the thecal sac and courses toward the neural foramen it is covered by arachnoid and the intervening subarachnoid space. The continuation of the arachnoid at and beyond the level of the dorsal root ganglion is termed the perineurium.[6] Typically, water-soluble myelographic contrast fills the subarachnoid space to the level of the dorsal root ganglion. A potential space is present beneath the perineurium; this space usually does not fill with contrast. CNRSD occurs when there is enlargement of the subarachnoid space that surrounds the exiting nerve root. The etiology is not known but may be related to increased hydrostatic pressure of the cerebrospinal fluid. Dilatation of the nerve root sleeve is located proximal to the dorsal root ganglion, permitting intrathecal contrast to communicate freely with the cystic structure during myelography.[4,6] Dilatation may be tubular or saccular,[2] the latter sometimes being termed a meningeal diverticulum.[4] CNRSD may be seen in as many as 18% of lumbar myelograms performed for low back pain or sciatica but is usually not a cause of symptoms.[2]

CNRSD differs from a Tarlov (perineurial) cyst, which occurs at the level of the dorsal root ganglion and contains neural elements in its wall.[6] A perineurial cyst may cause symptoms, which are relieved by surgery. The radiographic differentiation of CNRSD from perineurial cyst is difficult. Tarlov, who studied perineurial cysts with Pantopaque myelography, found a lack of filling of the cyst during the initial myelogram, although delayed myelographic filling was demonstrated in some cases.[6] Thus when Pantopaque contrast was used the perineurial cyst did not fill on the initial study, whereas CNRSD (or meningeal diverticulum) filled readily. This radiographic feature was used to differentiate these entities.[6] With the advent of water-soluble contrast this differentiation may no longer be valid. In one report, water-soluble contrast filled sacral perineurial cysts on the initial myelographic and CTM studies.[5] Filling of perineurial cysts with contrast is thought to be due to communication between the subarachnoid space and the potential space beneath the perineurium.[6]

CNRSD is demonstrated by CT as a round mass in the region of the neural foramen isodense with the thecal sac. This causes asymmetry of the perineural fat and may cause enlargement of the neural foramen. Scalloping of the adjacent vertebral body, pedicle, or pedicular-laminar junction may occur.[1,3,4] Multilevel involvement and dural ectasia are other features of this entity.[2,4] The most frequent sites of CNRSD are S1 and S2.[2] When intrathecal contrast is introduced, filling of the dilated nerve root sleeve is readily apparent. The perineurial cyst has a CT appearance similar to that of CNRSD, with a low-density cyst oc-

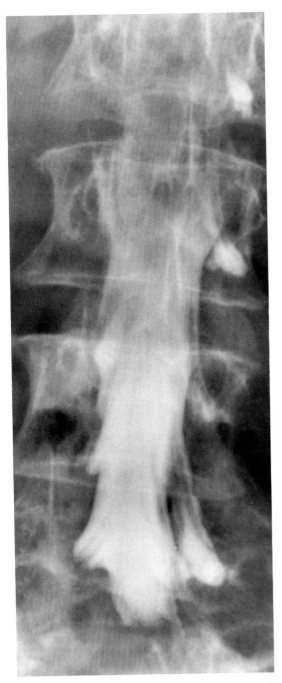

FIG. 8B. Cystic nerve root sleeve dilatation. Same patient as in Fig. 8A-1. Myelogram performed with water-soluble contrast demonstrates CNRSD at multiple lumbar levels.

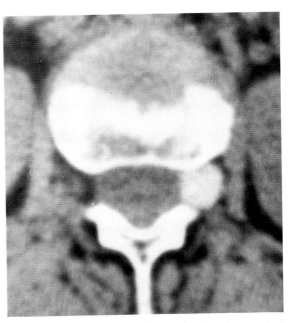

FIG. 8C. Cystic nerve root sleeve dilatation. Same patient as in Figs. 8A-1 and 8B. CTM performed several hours after myelogram. Contrast remains within the dilated nerve root sleeve and demonstrates its smooth, rounded configuration.

curring most frequently in the sacral levels.[7] These sacral cysts can attain large size, causing sacral canal enlargement and osseous erosion[5] (Figs. 8D, 8E). In the cervical region diverticula are seen as smooth, round, low-density structures that opacify readily when intrathecal contrast is introduced for a myelogram or CTM (Fig. 8F).

The CT appearance of CNRSD should permit differentiation from a lateral or extruded disc herniation, both of which typically have much higher density measurements. Conjoined nerve roots may be isodense with the thecal sac; however, this anomaly usually occurs at one level, is typically unilateral, and more often appears oval or elliptical rather than round when studied in the axial plane. Neurofibroma should be considered in the differential diagnosis of a perineurial cyst. Usually a neurofibroma has slightly higher CT density measurements than a cyst; however, when a neurofibroma degenerates it may have similar density measurements.[7] A neurofibroma can also cause osseous scalloping, thus further simulating a perineurial cyst. In general the CT findings described should permit differentiation of CNRSD or perineurial cyst from disc herniation or tumor; however, myelography or CTM can be performed if further evaluation is needed.

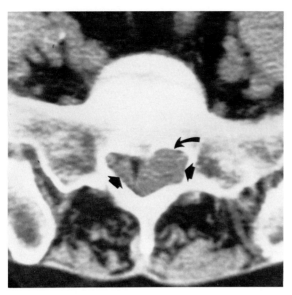

FIG. 8D. Sacral perineurial cysts. Axial CT of the superior sacrum. There are bilateral perineurial cysts *(straight arrows)*. The larger cyst on the left is causing pressure erosion of the posterior aspect of the sacrum *(curved arrow)*.

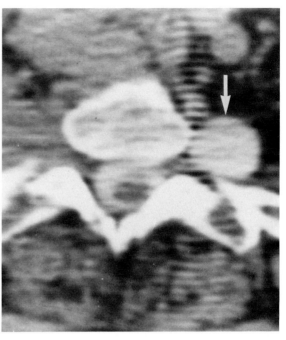

FIG. 8F. Cervical meningeal diverticulum. CTM demonstrates a round, contrast-filled diverticulum with smooth margins *(arrow)*.

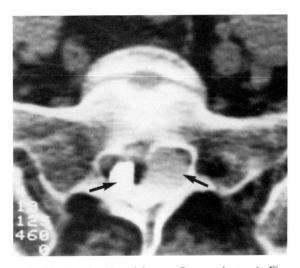

FIG. 8E. Sacral perineurial cysts. Same patient as in Fig. 8D. CTM reveals water-soluble contrast filling the large bilateral sacral cysts *(arrows)*.

References

1. Eisenberg D, Gomori JM, Findler G, et al: Symptomatic diverticulum of the sacral nerve root sheath: Case note. *Neuroradiology* 1985;27:183.
2. Larsen JL, Smith D, Fossan G: Arachnoidal diverticula and cystlike dilatations of the nerve-root sheaths in lumbar myelography. *Acta Radiol Diagn* 1980;21:141–145.
3. Naidich TP, McLone DG, Harwood-Nash DC: Arachnoid cysts, paravertebral meningoceles, and perineurial cysts, in Newton TH, Potts DG (eds): *Computed Tomography of the Spine and Spinal Cord.* San Anselmo, Calif. Clavadel Press, 1983, pp 383–396.
4. Neave VCD, Wycoff RR: Computed tomography of cystic nerve root sleeve dilatation: Case report. *J Comput Assist Tomogr* 1983;7:881–885.
5. Siqueira EB, Schaffer L, Kranzler LI, et al: CT characteristics of sacral perineurial cysts: Report of two cases. *J Neurosurg* 1984;61:596–598.
6. Tarlov IM: Spinal perineurial and meningeal cysts. *J Neurol Neurosurg Psychiatry* 1970;33:833–843.
7. Willinsky RA, Fazl M: Computed tomography of a sacral perineurial cyst: Case report. *J Comput Assist Tomogr* 1985;9:599–601.

CASE 9

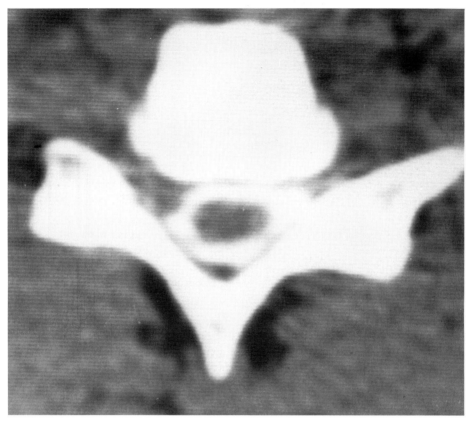

FIG. 9A. Axial CT at C6-C7 performed several hours after myelography with water-soluble contrast. This 50-year-old had pain radiating into the posterior aspect of the right arm and into the right thumb and index finger.

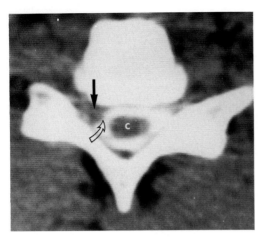

FIG. 9A-1. A lateral disc herniation *(straight arrow)*, has extruded into the right neural foramen. The disc herniation is causing flattening of the anterolateral portion of the subarachnoid space and truncation of the nerve root sleeve *(curved arrow)*. Compare with the normal left side. The spinal cord *(C)* is not compressed or displaced.

Cervical Disc Herniation

A lateral disc herniation is present on the right and extends into the neural foramen (Fig. 9A-1). There is asymmetry of the metrizamide-filled sac with flattening and compression of the right anterolateral portion of the subarachnoid space. At surgery, a free fragment of extruded disc was found in the right neural foramen.

Although CT scanning has become widely accepted in the diagnosis of lumbar disc disease, there has been slower acceptance of CT's ability to diagnose cervical disc disease. This is at least in part due to several factors inherent in the cervical spine examination:[7] First, the cervical spine lacks the abundant epidural fat present in the lumbar region. Disc herniation must be recognized as contrasted to the adjacent CSF and dural sac, which are more dense than fat. Although CSF and the dural sac are of lower density than disc, the density differences are more subtle than the disc/epidural fat interface of the lumbar levels. Second, the disc spaces are narrower, averaging 3 to 5 mm in height compared to approximately 8 mm in the lumbar levels. Third, the thin slices required for optimal visualization of disc herniation in turn lead to decreased contrast resolution. Fourth, artifact may be present at the lower cervical levels due to scanning through the plane of the shoulders.

Despite these factors, CT has been found to be an accurate method of diagnosing disc disease in the cervical spine.[1,3,5] It may be more accurate than myelography in evaluation of patients with cervical radiculopathy (neurologic deficit due to nerve root compression).[1,3-5] When a patient has a radiculopathy with a well-defined clinical level, soft disc herniation and spondylosis are diagnosed by CT at least as effectively as by myelography and with less risk and morbidity to the patient.[5] On the other hand, cervical myelopathy (neurologic deficit due to spinal cord compression) is more appropriately studied by myelography than by plain CT because of the extensive portion of the cervical canal that must be examined.[5,10,11] CTM may follow the myelogram to better assess cord compression and to differentiate disc herniation from spondylosis.[11] In one CT study, good correlation was found between the degree of spinal cord deformity due to disc disease or spondylosis and the results of surgery.[13] All patients having severe deformity and most patients with moderate deformity experienced substantial improvement after surgery.

The CT examination of the cervical spine for possible disc herniation begins with a lateral digital radiograph, which is required for precise localization. Thin, 1.5- to 2-mm slices should be obtained parallel to the disc.[3-5,12] Each level to be examined is evaluated with approximately seven contiguous slices at 1.5- to 2-mm intervals with an equal number of slices on either side of the disc. The patient is asked not to breathe, move, or swallow during the CT exposures. Magnification is utilized, and scans are photographed at both soft-tissue (e.g., window width 250 HU for cervical spine) and bone window (e.g., window width

1,000 HU) settings. Cervical and shoulder traction can be used to avoid shoulder artifact.[3]

There is no general agreement concerning the merits of unenhanced and enhanced CT study in the evaluation of cervical disc disease. Some authors routinely perform this examination without intrathecal or intravenous contrast.[3,4] We prefer, as do others,[1,5] to perform the CT study after introduction of intrathecal water-soluble contrast. One report states that unenhanced CT produces suboptimal visualization of the disc space in 50% of cases.[6] Intrathecal contrast enhances the visualization of the subarachnoid space, nerve roots, and spinal cord and permits better appreciation of spinal cord and nerve root compression.[11] In another report, CT studies performed either with or without intrathecal metrizamide were more accurate than myelography in diagnosing the cause of cervical radiculopathy.[5] Still other investigators prefer to use intravenous iodinated contrast-enhanced CT as their primary diagnostic procedure in cases of suspected cervical disc disease.[2,12] Intravenous contrast given in a high-volume bolus followed by drip infusion enhances the CT appearance of the epidural veins and dura. This increases the diagnostic accuracy of herniated nucleus pulposus by accentuating the interface between the subarachnoid space and the disc.[10] Intravenous contrast-enhanced CT also permits excellent visualization of the exiting cervical nerve root by opacifying the surrounding intervertebral venous plexus.[8] In some patients with cervical radiculopathy, the unenhanced CT scan may be equivocal or nondiagnostic; however, cervical disc herniation or extrusion may be clearly visualized after intravenous contrast enhancement of the dura and venous structures.[10] When the primary concern is evaluation of the spinal cord rather than the disc, intrathecal water soluble contrast is recommended. The CT study may be performed approximately 2 to 4 hours after the myelogram. When one is evaluating the spinal cord, thicker (4 to 5 mm) slices can be used.[9,12]

Most cervical disc herniations occur posterolaterally or laterally, causing compression of a nerve root while sparing the spinal cord. This may cause neck pain and a radiculopathy with sensory disturbance. The unenhanced CT examination reveals focal protrusion of the disc margin (Fig. 9B). With CTM there is compression and focal deformity of the contrast-filled subarachnoid space and truncation of the nerve root sleeve.[12] A far lateral disc extrusion may be seen as a soft-tissue mass in the neural foramen. Midline herniations tend to displace and compress the entire ventral surface of the thecal sac and, if large enough, compress the spinal cord (Fig. 9C). Central disc herniations are associated with less specific neurologic findings such as neck pain or intermittent signs of a radiculopathy or myelopathy.[12] CT can readily differentiate spondylosis from soft disc herniation (Fig. 9D).

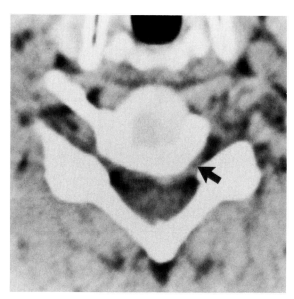

FIG. 9B. Cervical disc herniation. CT at C5-C6 without intrathecal or intravenous contrast demonstrates disc herniation *(arrow)* on the left side extending posterolaterally and laterally into the the neural foramen.

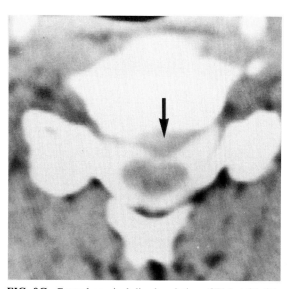

FIG. 9C. Central cervical disc herniation. CTM at C4-C5. There is central disc herniation *(arrow)* compressing the contrast-filled subarachnoid space anteriorly. In the cervical spine, intrathecal contrast is helpful in delineating the subarachnoid space, nerve roots, and spinal cord.

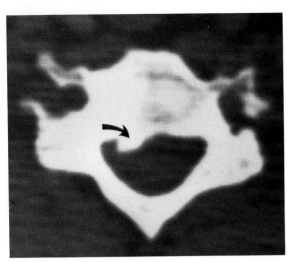

FIG. 9D. Cervical osteophyte. This patient has an osteophyte *(arrow)* on the right at C5-C6. CT can readily distinguish an osteophyte from a soft disc herniation.

References

1. Badami JP, Norman D, Barbaro NM, et al: Metrizamide CT myelography in cervical myelopathy and radiculopathy: Correlation with conventional myelography and surgical findings. *AJNR* 1985;6:59–64, *AJR* 1985;144:675–680.
2. Balériaux D, Noterman J, Ticket L: Recognition of cervical soft disk herniation by contrast-enhanced CT. *AJNR* 1983;4:607–608.
3. Coin CG: Computed tomography of cervical disc disease (herniation and degeneration), in Post MJD (ed): *Computed Tomography of the Spine*. Baltimore, Williams & Wilkins, 1984, pp 387–405.
4. Coin CG, Coin JT: Computed tomography of cervical disk disease: Technical considerations with representative case reports. *J Comput Assist Tomogr* 1981;5:275–280.
5. Daniels DL, Grogan JP, Johansen JG, et al: Cervical radiculopathy: Computed tomography and myelography compared. *Radiology* 1984;151:109–113.
6. Dublin AB, McGahan JP, Reid MH: The value of computed tomographic metrizamide myelography in the neuroradiological evalution of the spine. *Radiology* 1983;146:79–86.
7. Haughton VM, Williams AL. *Computed Tomography of the Spine*. St Louis, CV Mosby, 1982.
8. Heinz ER, Yeates A, Burger P, et al: Opacification of epidural venous plexus and dura in evaluation of cervical nerve roots: CT technique. *AJNR* 1984;5:621–624.
9. Orrison WW, Johansen JG, Eldevik OP, et al: Optimal computed-tomographic techniques for cervical spine imaging. *Radiology* 1982;144:180–182.
10. Russell EJ, D'Angelo CM, Zimmerman RD, et al: Cervical disk herniation: CT demonstration after contrast enhancement. *Radiology* 1984:152:703–712.
11. Scotti G, Scialfa G, Pieralli S, et al: Myelopathy and radiculopathy due to cervical spondylosis: Myelographic-CT correlations. *AJNR* 1983;4:601–603.
12. Yeates AE: Computed tomographic evaluation of cervical pain syndromes, in Genant HK (ed): *Spine Update 1984: Perspective in Radiology, Orthopaedic Surgery, and Neurosurgery*. San Francisco, Radiology Research and Education Foundation, 1983, pp 291–307.
13. Yu YL, Stevens JM, Kendall B, et al: Cord shape and measurements in cervical spondylotic myelopathy and radiculopathy. *AJNR* 1983;4:839–842.

CASE 10

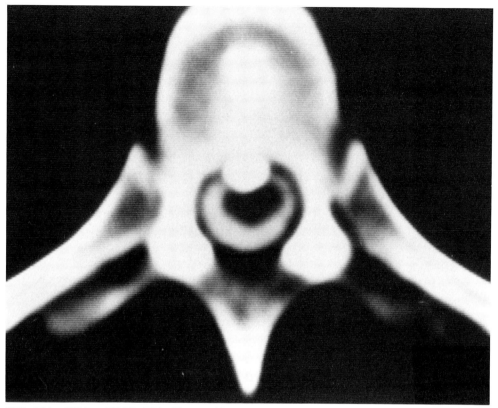

FIG. 10A. CTM at T6-T7. This 42-year-old woman had pain in the left side of the thorax and leg weakness.

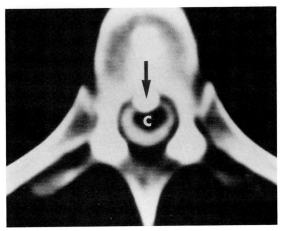

FIG. 10A-1. Thoracic disc herniation. Densely calcified disc herniation *(arrow)* compresses the contrast-filled subarachnoid space and spinal cord *(C)*.

Thoracic Disc Herniation

There is a densely calcified "mass" in the anterior epidural space causing compression of the subarachnoid space and spinal cord (Fig. 10A-1). This is a thoracic disc herniation, which is rare, occurring much less frequently than lumbar or cervical disc herniation. Trauma, which may be mild, has been noted in about 25% of cases.[1] Patients with thoracic disc herniation most frequently present with back pain and less often with pain in the legs, groin, or pelvis. Marked numbness and paresthesias below the level of the herniation may suggest the diagnosis clinically.[3] When large protrusions occur at the lower thoracic levels, compression of the conus medullaris and cauda equina may lead to the cauda equina syndrome.

Herniation of a thoracic disc may occur at any level but is most frequent in the lower thoracic region. Approximately 75% of thoracic disc herniations occur below T8, with T11-T12 the most frequent site followed by T10-T11 and T9-T10.[1] Conventional radiography demonstrates calcification within the spinal canal at the level of the intervertebral disc space narrowing in approximately 55% of cases of thoracic disc herniation.[5] Several cases of thoracic disc herniation demonstrated by CT have been reported.[1,2,4,6,7] Calcification seen within the canal on conventional radiography can direct the CT examination to a specific level for study. A noncalcified thoracic disc herniation is less easily demonstrated by unenhanced CT for two significant reasons: First, it is difficult to determine the clinical level to be examined because the sensory deficit that the patient experiences may extend above the dermatomal level corresponding to the anatomical lesion.[3] Second, the amount of epidural fat in the thoracic spine is variable and, when sparse, may lead to difficulty in identifying soft disc herniation. Myelography plays a primary role in diagnosis of thoracic disc herniation. An extradural impression is seen myelographically; however, clinical correlation is important because asymptomatic thoracic defects are not uncommon. CTM may follow myelography to confirm the diagnosis of a thoracic disc herniation. In some cases myelographic examination may fail to demonstrate a thoracic disc herniation, which is then correctly diagnosed with CTM.[1] Rarely, more than one thoracic herniated nucleus pulposus can be demonstrated by CTM.[2]

References

1. Arce CA, Dohrmann GJ: Thoracic disc herniation: Improved diagnosis with computed tomographic scanning and a review of the literature. *Surg Neurol* 1985;23:356–361.
2. Bhole R, Gilmer RE: Two-level thoracic disc herniation. *Clin Orthop* 1984;190:129–131.
3. Carson J, Gumpert J, Jefferson A: Diagnosis and treatment of thoracic intervertebral disc protrusions. *J Neurol Neurosurg Psychiatry* 1971;34:68–77.
4. Hochman MS, Pena C, Ramirez R: Calcified herniated thoracic disc diagnosed by computerized tomography: Case report. *J Neurosurg* 1980;52:722–723.
5. McAllister VL, Sage MR: The radiology of thoracic disc protrusion. *Clin Radiol* 1976;27:291–299.
6. Schimel S, Deeb ZL: Herniated thoracic intervertebral disks. *CT* 1985;9:141–143.
7. van Ameyden van Duym FC, van Wiechen PJ: Herniation of calcified nucleus pulposus in the thoracic spine: Case report. *J Comput Assist Tomogr* 1983;7:1122–1123.

CASE 11

FIG. 11A. Axial CT scan through the plane of the odontoid process of C2. The study was performed to evaluate for possible recurrent tumor. What can be done to examine this patient more adequately with CT?

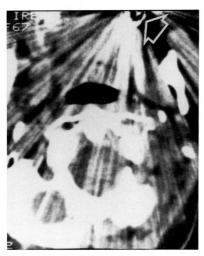

FIG. 11A-1. Dental filling artifact. There is extensive streaking artifact due to a dental filling *(arrow)*. The artifact prevents adequate evaluation of the spinal canal.

Dental Filling Artifact

Streaking artifact is traversing the spinal canal and degrading the image to an unacceptable level (Fig. 11A-1). This artifact is caused by a metallic dental filling within a tooth and can be eliminated if identified during the examination. A lateral digital radiograph is generated, and a different plane of scanning is selected to avoid the dental filling (Fig. 11B). Transaxial scans now demonstrate the same area of the spinal canal free of artifact (Fig. 11C). In this case an epidural tumor is identified.

The referring and imaging physicians should be familiar with the many factors that cause image deg-

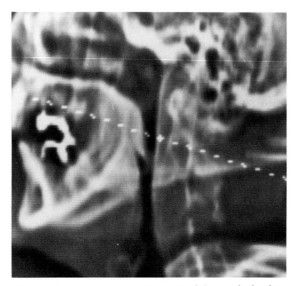

FIG. 11B. Lateral digital radiograph of the cervical spine. Same patient as in Fig. 11A-1. A new plane of scanning has been selected as shown by the position of the cursor line, which is angled to avoid the dental fillings. Subsequent axial scans can then be obtained at and above the plane of the cursor line.

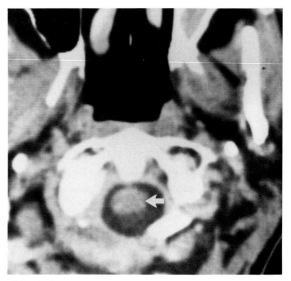

FIG. 11C. Axial CT scan free of the artifact that was present in Fig. 11A-1. This CT was obtained after altering the plane of scanning as shown in Fig. 11B. Approximately the same level of spinal cord is studied in Figs. 11A-1 and 11C, although the posterior bony structures differ because of changes in gantry angulation. This CT was obtained after intravenous contrast enhancement and demonstrates a contrast-enhanced recurrent melanoma *(arrow)*.

radation. In the lower cervical spine, especially the C7-T1 level, images may be suboptimal because of streak artifact created by scanning through the plane of the shoulders (Figs. 11D, 11E). This artifact may occur even though cervical and shoulder traction are used with the patient stretching the arms downward as far as possible.

The examining table has a patient weight restriction, which varies among the different manufacturers but may be, for example, 300 pounds. A patient weighing more than the weight restriction should not be examined. However, obese patients whose weight is below the critical level may still not be satisfactory candidates for study. The scans of obese patients are often too grainy and artifact laden for adequate evaluation. There are CT equipment refinements that permit a higher milliampere-second technique to obtain a satisfactory scan. However, this

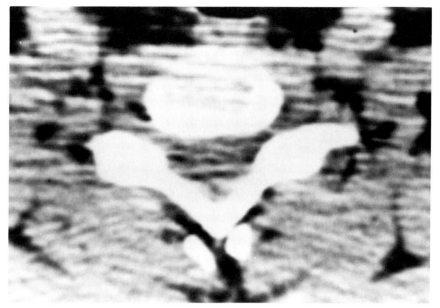

FIG. 11D. Streak artifact. Axial CT at C6-C7. The artifact prevents optimal evaluation of the structures within the spinal canal. This type of artifact may be seen in the lower cervical spine and in the lumbar spine of obese patients.

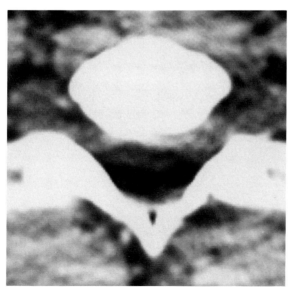

FIG. 11E. "Shoulder" artifact. Axial CT at C7-T1 distorted by low-density artifact within the spinal canal created by scanning through the plane of the shoulders. This type of artifact limits the ability to diagnose abnormalities within the spinal canal such as disc herniation.

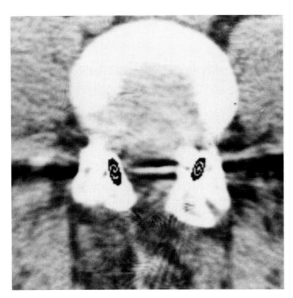

FIG. 11F. Surgical screw artifact. Bilateral metal screws traverse the articular processes. Artifact occurs because of the presence of the screws and degrades the image within the spinal canal.

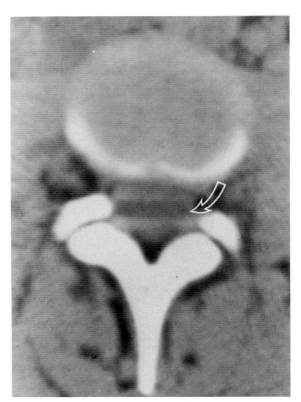

FIG. 11G. Facet artifact. Axial CT at L3-L4. There is a low-density linear artifact *(arrow)* traversing the spinal canal horizontally between the facet joints. This facet artifact is caused by the sudden change in density between the osseous articular processes and the intervening facet joints as measured by the CT scanner.

technique increases radiation dosage substantially and should be used prudently.

In addition to tooth filling artifacts, sources of metal density within or adjacent to the spinal canal may prevent optimal or even adequate examination. Pantopaque contrast from a previous myelogram, surgical clips, surgical screws, and bullet fragments are other causes of artifact (Fig. 11F). The so-called facet artifact is a horizontal low-density line that may be seen within the spinal canal traversing the plane between the anterior medial extension of both facet joints (Fig. 11G). This artifact is due to the edge effect caused by a sudden change in density of the surrounding structures as measured by the scanner. The extreme difference in density between the bone of the articular processes and the intervening joint space causes this low-density artifact in some patients. The relatively higher density area within the spinal canal seen anterior to the artifact should not be mistaken for disc herniation.

CASE 12

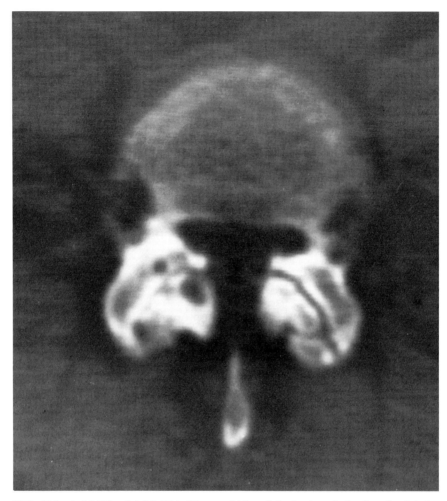

FIG. 12A. Axial CT at L4-L5. This 67-year-old female had pain radiating into the right leg.

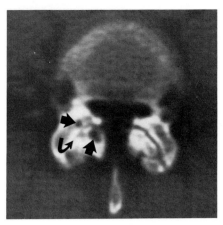

FIG. 12A-1. Osteoarthritis of the facet joints. There is severe narrowing of the right facet joint *(curved arrow)* and moderate narrowing of the left facet joint. Subchondral erosions and cysts *(straight arrows)* and sclerosis are present. This image was photographed at bone window settings (window width, 1,024 HU; window level, 250 HU).

Osteoarthritis of Facet Joints

There is severe osteoarthritis of the facet joints with joint space narrowing, subchondral erosions, and sclerosis (Fig. 12A-1). The facet joints are lined with synovial membrane and are formed by articulation of the inferior articular process of the vertebra above and the superior articular process of the vertebra below. CT is an ideal method of evaluating facet joints of the lumbar spine because of the oblique orientation of the joints. The orientation of the lumbar facet joints varies with the level examined. The facet joints tend toward a more sagittal orientation at L3-L4, a more coronal plane at L5-S1, and an intermediate position at L4-L5.[9] Facet joints are best studied at bone window settings (e.g., window width 1,000 to 2,000 HU and window level 200 to 350 HU) rather than the usual soft-tissue window settings used to visualize the lumbar disc (e.g., window width 500 HU and window level 25 to 50 HU).

A normal facet joint has smooth, regular cortical margins with a joint space width of 2.0 to 4.0 mm (Fig. 12B). Abnormal facet joints are frequently discovered with CT. In one study, patients with low back pain and/or sciatica were studied by CT, and 43% had abnormal facet joints whereas only 18% had disc herniation.[3] The most frequent abnormality of facet joints is an osteophyte, an outgrowth of cortical bone derived from the articular margin and therefore lacking a medullary space (Fig. 12C). Articular facet hypertrophy is another common finding and appears as an enlargement of the articular process with normal medullary and cortical proportions[2,3] (Fig. 12D). Other CT findings of degenerative arthritis of facet joints include joint space narrowing (<2 mm), subchondral

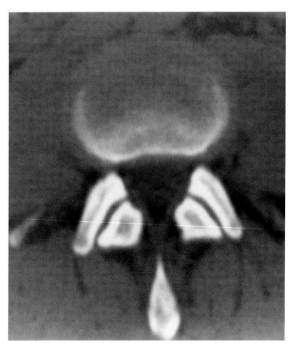

FIG. 12B. Normal facet joints. The normal articular processes have smooth regular cortical margins. The normal facet joint measures 2 to 4 mm in width. The facet joints are ideally studied at bone window settings.

sclerosis, and subchondral erosions and cysts[2,3] (Figs. 12A-1, 12E). Calcifications in the periarticular region and gas within the facet joint (vacuum facet phenomenon) may also be seen. Subluxation of facet joints may be present in association with degenerative spondylolisthesis (anterior subluxation of inferior articular processes) or retrospondylolisthesis (posterior subluxation of inferior articular processes) (Fig. 12F).

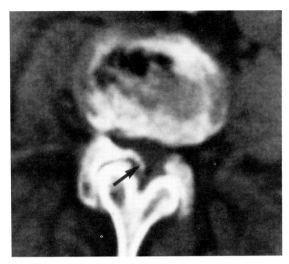

FIG. 12C. Osteophyte. Axial CT demonstrates an osteophyte *(arrow)* derived from the superior articular process on the right causing spinal stenosis. The facet joints are narrow.

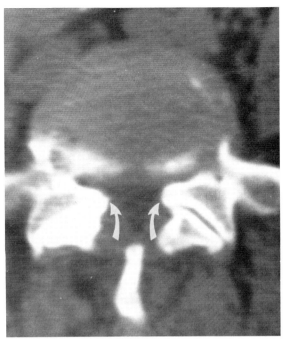

FIG. 12D. Hypertrophy of articular processes. CT at L5-S1 with bilateral hypertrophy of the superior articular processes of S1 *(arrows)*. The right facet joint is narrow.

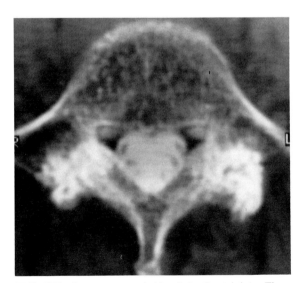

FIG. 12E. Severe osteoarthritis of the facet joints. There is extensive subchondral sclerosis, hypertrophy, and posterior osteophyte formation as well as obliteration of the facet joints. This CT scan was obtained at L5 after introduction of intrathecal water-soluble contrast. Note the lack of stenosis despite severe osteoarthritis.

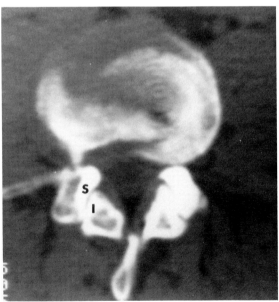

FIG. 12F. Retrospondylolisthesis at L3-L4. There is posterior subluxation of the inferior articular process of L3 *(I)* in relation to the superior articular process of L4 *(S)*.

Each facet joint has a dual innervation. The joint capsule of the superior articular process is innervated by branches of the dorsal ramus of the spinal nerve at the same level, whereas the joint capsule of the inferior articular process is innervated from branches of the dorsal ramus of the spinal nerve of the next higher vertebral level.[4] Pain arising from a facet joint can be referred to the structures innervated by both dorsal root ganglia involved.[2,4] Patients with lumbar facet arthropathy may have symptoms of low back pain and sciatica simulating disc herniation. Several methods of treating patients clinically suspected of having the facet joint syndrome have been recommended and include fluoroscopically guided intra-articular injection of a local anesthetic and steroid suspension,[1,3,4,6] percutaneous radiofrequency facet denervation,[8] and surgical fusion.[7] Although presently there is no definite way to identify which facet joints might be responsible for significant low back pain,[7] CT may play a role in this evaluation. In a group of patients who had CT examination prior to facet block injection, almost all who obtained relief following the injection had abnormalities of facet joints demonstrated by CT.[3] Some patients with abnormal facet joints had no pain relief, suggesting that their pain was either derived from another source or was at least not originating from the capsular innervation.[3] Large osteophytes of articular processes, for example, may cause central or lateral stenosis and be responsible for a radiculopathy that would not be relieved by intra-articular injection. Some of these patients might instead be successfully treated by foraminotomy. In the evaluation of patients for possible facet joint syndrome, CT can diagnose other causes of low back pain and sciatica such as herniated nucleus pulposus, spinal stenosis, and metastasis.[5]

References

1. Carrera GF: Lumbar facet joint injection in low back pain and sciatica: Preliminary results. *Radiology* 1980;137:665–667.
2. Carrera GF, Haughton VM, Syvertsen A, et al: Computed tomography of the lumbar facet joints. *Radiology* 1980;134:145–148.
3. Carrera GF, Williams AL: Current concepts in evaluation of the lumbar facet joints. *CRC Crit Rev Diagn Imaging* 1985;21:85–104.
4. Destouet JM, Gilula LA, Murphy WA, et al: Lumbar facet joint injection, technique, clinical correlation, and preliminary results. *Radiology* 1982;145:321–325.
5. Federle MP, Moss AA, Margolin FR: Role of computed tomography in patients with "sciatica." *J Comput Assist Tomogr* 1980;4:335–341.
6. Mooney V, Robertson J: The facet syndrome. *Clin Orthop* 1976;115:149–156.
7. Raymond J, Dumas J-M: Intraarticular facet block: Diagnostic test or therapeutic procedure? *Radiology* 1984;151:333–336.
8. Shealy CN: Facet denervation in the management of back and sciatic pain. *Clin Orthop* 1976;115:157–164.
9. Van Schaik JPJ, Verbiest H, Van Schaik FDJ: The orientation of the laminae and facet joints in the lower lumbar spine. *Spine* 1985;10:59–63.

CASE 13

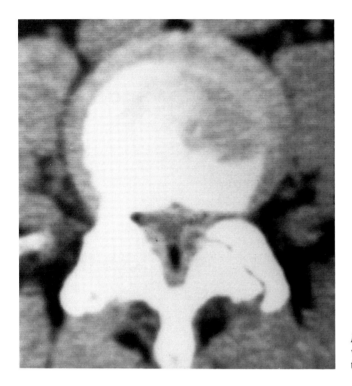

FIG. 13A. Axial CT at L4-L5 viewed at soft-tissue window settings. This 61-year-old male has symptoms of spinal stenosis.

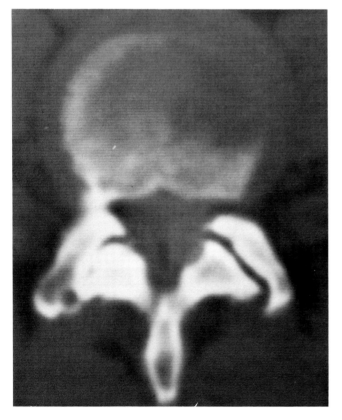

FIG. 13B. Same scan as in Fig. 13A viewed at bone window settings.

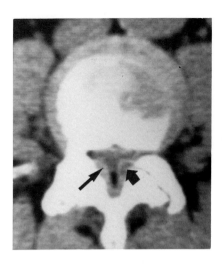 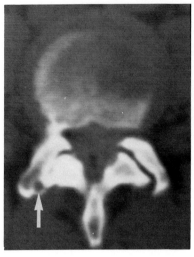

FIG. 13A-1. A low-density synovial cyst *(short arrow)* with a dense rim is present on the left side. There is thickening of the ligamentum flavum on the right *(long arrow)*. The synovial cyst and the prominent ligamentum flavum are causing spinal stenosis. Bulging of the disc is also present.

FIG. 13B-1. Bilateral osteoarthritis of the facet joints. There is joint space narrowing and a subchondral cyst *(arrow)* on the right. Hypertrophic changes of the articular processes are seen on both sides.

Synovial Cyst

There is a low-density cystic mass in the epidural space, surrounded by a dense rim. The cystic structure is located adjacent to the medial aspect of the left facet joint at the L4-L5 level (Fig. 13A-1). Degenerative disease of the facet joint can be seen when the CT scan is viewed at bone window settings (Fig. 13B-1). These are the typical findings of a synovial cyst associated with degenerative arthritis of the facet joint. Synovial cysts of the lumbar spine are thought to occur in response to degenerative changes of the synovial membrane-lined facet joints. Joint effusion bulges the joint capsule, and the synovial membrane herniates through the joint capsule-ligamentum flavum complex to form a synovial cyst. Almost all reported synovial cysts have occurred at L4-L5, possibly because of the increased motion that occurs at this level.[1,3,4]

Nonspecific radiographic and myelographic findings have been described associated with synovial cysts. Conventional radiography may reveal changes of degenerative spondylolisthesis, and myelography may demonstrate posterolateral extradural impressions.[1] However, the CT findings shown in this case are characteristic of a synovial cyst. Typically, a cystic structure is located adjacent to a degenerated facet joint at the L4-L5 level. The cyst is surrounded by a rim of calcification.[2–4] Gas may be detected within the cyst, having dissected from an adjacent facet joint with a vacuum phenomenon.[5,6]

Occasionally a synovial cyst may cause nerve root compression with radiculopathy, requiring surgical excision for cure.[2,4] More frequently, however, symptoms are thought to be related to facet joint arthropathy rather than pressure effect from the cyst.[3] Surgical intervention may not be required since these cysts often regress spontaneously and may not be identified on repeat CT examination or during surgery. The characteristic CT findings of synovial cyst should help distinguish this entity from a free disc fragment or an epidural neoplasm.

References

1. Bhushan C, Hodges FJ III, Wityk JJ: Synovial cyst (ganglion) of the lumbar spine simulating extradural mass. *Neuroradiology* 1979;18:263–268.
2. Casselman ES: Radiologic recognition of symptomatic spinal synovial cysts. *AJNR* 1985;6:971–973.
3. Hemminghytt S, Daniels DL, Williams AL, et al: Intraspinal synovial cysts: Natural history and diagnosis by CT. *Radiology* 1982;145:375–376.
4. Kurz LT, Garfin SR, Unger AS, et al: Intraspinal synovial cyst causing sciatica. *J Bone Joint Surg Am* 1985;67-A:865–871.
5. Schulz EE, West WL, Hinshaw DB, et al: Gas in a lumbar extradural juxtaarticular cyst: Sign of synovial origin. *AJR* 1984;143:875–876.
6. Spencer RR, Jahnke RW, Hardy TL: Dissection of gas into an intraspinal synovial cyst from contiguous vacuum facet. *J Comput Assist Tomogr* 1983;7:886–888.

CASE 14

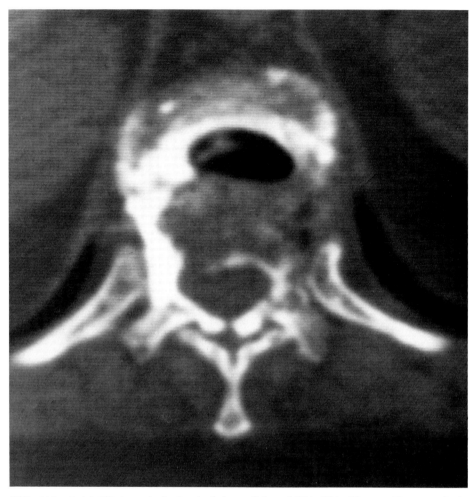

FIG. 14A. Axial CT through the level of the pedicles of T11. This 55-year-old woman had a 2-month history of back pain.

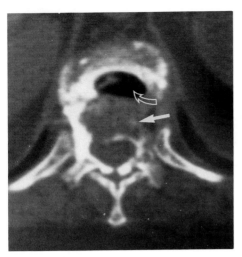

FIG. 14A-1. Metastasis with intraosseous gas. There is extensive osteolytic metastasis *(straight arrow)* to the vertebral body and left pedicle from carcinoma of the breast. Epidural extension of tumor is present. In addition, intraosseous gas *(curved arrow)* can be seen within the vertebral body. The gas has negative CT attenuation values, which make it readily identifiable.

Metastasis with Intraosseous Gas

This is a case of intraosseous gas associated with metastatic disease of the vertebral body (Fig. 14A-1). This patient has osteolytic destruction of the vertebral body secondary to carcinoma of the breast. Occasionally, as in this case, intraosseous gas may be present when vertebral collapse occurs as a result of primary or metastatic tumor.[13]

Gas is frequently demonstrated on CT examination, especially within discs and facet joints. Gas is readily identified on CT examination because of its negative attenuation values, which range from -100 to $-1,000$ HU, depending on partial volume averaging of surrounding structures. Patients with primary intervertebral osteochondrosis (degenerative disc dis-

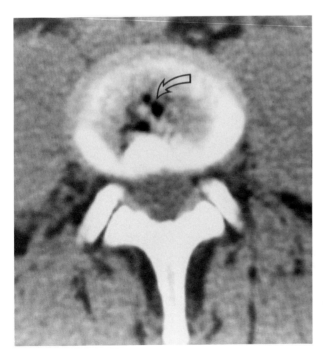

FIG. 14B. Vacuum disc phenomenon. Gas *(arrow)* is present within the disc. This abnormality is frequently identified by CT in patients with degenerative disc disease.

ease) commonly have gas within the disc (vacuum disc phenomenon) (Fig. 14B). With advancing age, clefts form in the nucleus pulposus and later progress to involve the fibers of the annulus fibrosus.[13] Gas is released from surrounding tissues and accumulates within the fissures of the disc.[6] This gas is approximately 90% nitrogen and 10% a combination of other gases.[6] The vacuum disc phenomenon can be identified with conventional radiography, especially when studied in spinal extension. However, the gas is more readily found with CT, being demonstrated in 50% of patients over 40 years of age and 75% of patients over 60 years.[9] A vacuum disc phenomenon may also be seen in patients with secondary intervertebral osteochondrosis associated with calcium pyrophosphate dihydrate crystal deposition disease, alkaptonuria, tumor, or trauma.[9,13,14]

The CT examination may demonstrate gas within an intravertebral disc herniation (Schmorl's node). As shown in Fig. 14A-1, intraosseous gas may occur in association with primary or metastatic tumors. Another cause of intraosseous gas is the intravertebral vacuum cleft, thought to represent a local ischemic vertebral fracture not associated with tumor, inflammation, or trauma.[11] This vertebral osseous necrosis may occur, for example, in patients receiving steroid medication.

Gas may be present within a facet joint (vacuum facet) (Fig. 14C). The gas appears as a linear or elliptical low-attenuation area within the facet joint and is noted in 20% of patients evaluated for back pain.[3] The significance of this finding is uncertain but may be related to a lax joint capsule, uneven apposition of the joint surfaces, or a normal joint undergoing distraction due to position.[3] The vacuum facet may be seen with other changes of severe osteoarthritis of the facet joints[9] and has also been described in association with degenerative spondylolisthesis.[10] A synovial cyst located adjacent to a vacuum facet may have gas within its center.[15]

Gas may be seen within the spinal canal after lumbar puncture or surgery. Gas may be free in the epidural space adjacent to a vacuum phenomenon of the disc or facet joint. It is suggested that the gas within the disc or facet joint escapes into the canal secondary to a ligamentous tear.[8] Gas within a herniated disc can also be detected with CT.[1,4,7,8,12] The surrounding soft tissue of the herniated disc can usually be identified to distinguish this from gas that is free in the canal (Fig. 14D). Occasionally a disc herniation may be present with CT findings of gas within the canal unaccompanied by CT demonstration of soft-tissue disc material.[5]

CT demonstration of other unusual causes of epidural gas has been described. Epidural gas may be found secondary to a pelvic inflammatory process with internal fistula.[2] Gaseous distension of the thecal sac has been described in a postoperative patient in whom

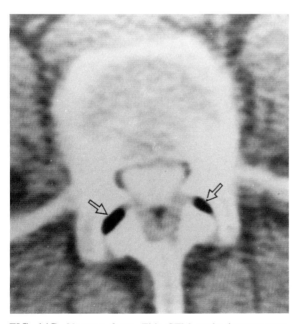

FIG. 14C. Vacuum facet. This CTM study demonstrates gas within both facet joints *(arrows)*. The gas appears as elliptical areas with negative CT attenuation.

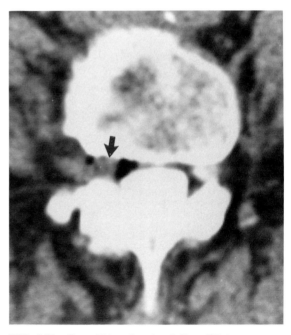

FIG. 14D. Gas within a herniated disc. CT demonstrates gas on the right within a disc herniation *(arrow)*. The disc herniation extends into the neural foramen. Note asymmetric contrast filling of the nerve root sleeves.

CT demonstration of the gas led to repair of an inadvertent surgical tear.[16] CT is a sensitive method of detecting gas in the spine and spinal canal, and evaluation of surrounding osseous and soft-tissue structures is helpful in determining the etiology and significance of the gas.

References

1. Austin RM, Bankoff MS, Carter BL: Gas collections in the spinal canal on computed tomography. *J Comput Assist Tomogr* 1981;5:522–524.
2. Burke V, Mall JC: Epidural gas: An unusual complication of Crohn disease. *AJNR* 1984;5:105–106.
3. Carrera GF, Haughton VM, Syvertsen A, et al: Computed tomography of the lumbar facet joints. *Radiology* 1980;134:145–148.
4. Dillon WP, Kaseff LG, Knackstedt VE, et al: Computed tomography and differential diagnosis of the extruded lumbar disc. *J Comput Assist Tomogr* 1983;7:969–975.
5. Elster AD, Jensen KM: Vacuum phenomenon within the cervical spinal canal: CT demonstration of a herniated disc. *J Comput Assist Tomogr* 1984;8:533–535.
6. Ford LT, Gilula LA, Murphy WA, et al: Analysis of gas in vacuum lumbar disc. *AJR* 1977;128:1056–1057.
7. Fries JW, Abodeely DA, Vijungco JG, et al: Computed tomography of herniated and extruded nucleus pulposus. *J Comput Assist Tomogr* 1982;6:874–887.
8. Gulati AN, Weinstein ZR: Gas in the spinal canal in association with the lumbosacral vacuum phenomenon: CT findings. *Neuroradiology* 1980;20:191–192.
9. Lardé D, Mathieu D, Frija J, et al: Spinal vacuum phenomenon: CT diagnosis and significance. *J Comput Assist Tomogr* 1982;6:671–676.
10. Lefkowitz DM, Quencer RM: Vacuum facet phenomenon: A computed tomographic sign of degenerative spondylolisthesis. *Radiology* 1982;144:562.
11. Maldague BE, Noel HM, Malghem J: The intravertebral vacuum cleft: A sign of ischemic vertebral collapse. *Radiology* 1978;129:23–29.
12. Orrison WW, Lilleas FG: CT demonstration of gas in a herniated nucleus pulposus. *J Comput Assist Tomogr* 1982;6:807–808.
13. Resnick D, Niwayama G, Guerra J Jr, et al: Spinal vacuum phenomena: Anatomical study and review. *Radiology* 1981;139:341–348.
14. Schabel SI, Moore TE, Rittenberg GM, et al: Vertebral vacuum phenomenon: A radiographic manifestation of metastatic malignancy. *Skeletal Radiol* 1979;4:154–156.
15. Spencer RR, Jahnke RW, Hardy TL: Dissection of gas into an intraspinal synovial cyst from contiguous vacuum facet. *J Comput Assist Tomogr* 1983;7:886–888.
16. Teplick JG, Haskin ME: Computed tomography of the postoperative lumbar spine. *AJNR* 1983;4:1053–1072, *AJR* 1983;141:865–884.

CASE 15

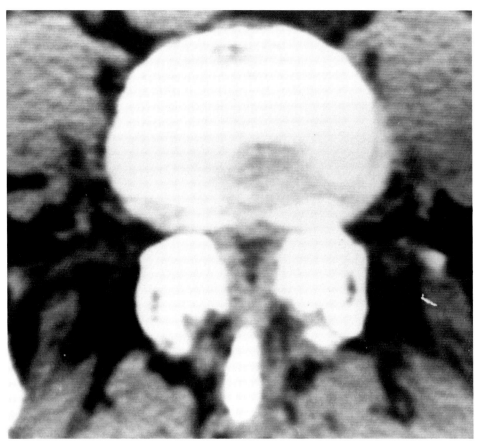

FIG. 15A. Axial CT at L4-L5 in a 59-year-old male with chronic back pain.

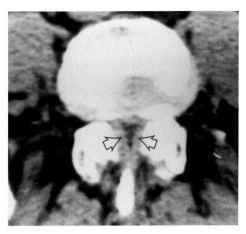

FIG. 15A-1. Central spinal stenosis. There is marked hypertrophy of the ligamenta flava *(arrows)* causing compression of the thecal sac. In addition there is mild bulging of the annulus and osteoarthritis of the facet joints.

Central Spinal Stenosis

There is degenerative central spinal stenosis secondary to marked thickening of the ligamenta flava (Fig. 15A-1). Anatomic classification divides spinal stenosis into central stenosis with impingement on the cauda equina (or the spinal cord at the cervical, thoracic, and upper lumbar levels) and lateral stenosis with impingement on the nerve root. Lateral stenosis can be further subdivided into lateral recess stenosis (subarticular) and neural foraminal stenosis (lateral nerve root canal).[7]

Another classification of spinal stenosis is congenital-developmental versus acquired stenosis. Congenital-developmental stenosis includes both idiopathic and achondroplastic stenosis, whereas acquired stenosis includes degenerative (both central and lateral), spondylolisthetic, iatrogenic (postsurgical fibrosis, fusion, chemonucleolysis), and posttraumatic stenosis.[1] Other causes such as Paget's disease and fluorosis are included among the acquired forms. Patients with congenital stenosis may develop degenerative changes and are then classified as having a combined form of stenosis.

Developmental and degenerative forms of spinal stenosis can be differentiated. Patients with developmental stenosis typically have uniform narrowing over several or all lumbar segments.[6] Examination at the level of the pedicles demonstrates a decreased AP diameter. There is a decreased interpedicular distance in approximately 20% of the vertebrae studied.[10] The pedicles may be short and the pedicles and laminae thickened in patients with developmental stenosis (Figs. 15B, 15C).

Degenerative spinal stenosis is typically segmental rather than uniform, with stenosis occurring at the level of the disc spaces and articular processes. Between these stenotic segments, the spinal canal and thecal sac may be normal in size.[6] Hypertrophic spurs derived from the inferior articular processes and less frequently from the superior articular processes may cause central spinal stenosis of the degenerative form (Figs. 15D, 15E). In many cases, marked thickening of the ligamentum flavum is the major cause of central stenosis (Fig. 15A-1). Frequently, generalized bulging of the disc, thickening of the ligamentum flavum, and hypertrophy of the articular processes act in combination to cause stenosis.

Patients with developmental spinal stenosis may be asymptomatic for years, their thecal sac just barely accommodated by the small canal. As minor degenerative changes occur, which would usually go unnoticed, these compromised patients may develop symptoms. This form of spinal stenosis is referred to as a combined form, having features of both developmental and degenerative stenosis. The entire canal may be narrowed, with additional segmental narrowing occurring at the level of the disc spaces and articular processes secondary to disc bulging and facet joint hypertrophy.

Central spinal stenosis can be diagnosed by both CT and myelography. In examining the CT scan for central stenosis we must study the bony canal, the thecal sac, and the epidural fat. CT determination of spinal canal size varies with the window width, window levels, and density of the intrathecal contents.[11] The most consistent measurements are obtained using a wide window width (1,000 to 4,000 HU using an extended scale). The window level is ideally set at the average between the CT numbers of the object

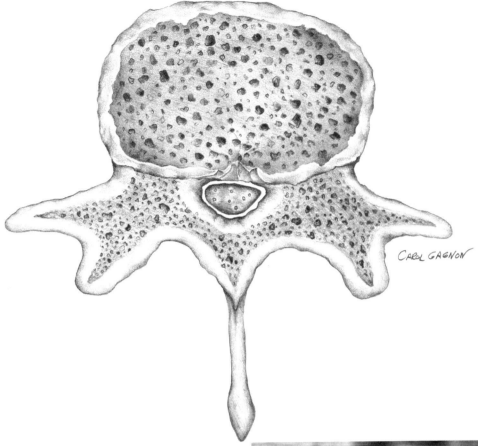

FIG. 15B. Developmental spinal stenosis. Axial view of a lumbar vertebra at the level of the pedicles. Note the short broad pedicles and thickened laminae, which lead to a decreased AP diameter of the canal.

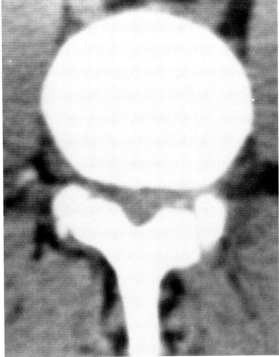

being measured and the surrounding structures as determined in Hounsfield units.[11] The use of intrathecal water-soluble contrast permits more reproducible canal size measurements.[11] The gantry angle does not have a significant effect on spinal canal measurement as long as scan slice thickness is 5 mm or less.[3] When scan slice thickness is 10 mm, gantry angulation that exceeds 15° from the transverse plane of the canal may result in underestimation of canal size.[3]

In the examination of patients with congenital-developmental lumbar stenosis, CT studies have suggested that a midsaggital diameter of 10 mm or less indicates absolute stenosis that may cause compression of the cauda equina.[12] Patients with AP diame-

FIG. 15C. Developmental central spinal stenosis. Axial CT just cephalad to the L4-L5 disc. The AP diameter of the canal is narrowed. Bilateral thickening of the laminae is seen.

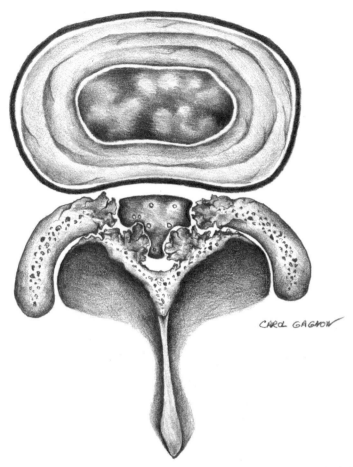

FIG. 15D. Degenerative central spinal stenosis. Axial view of a lumbar vertebra at the level of the disc. Note the osteophytes derived from the articular processes, which lead to compression of the thecal sac.

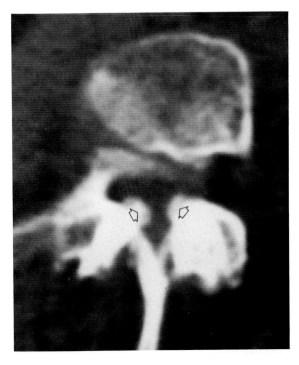

FIG. 15E. Degenerative central spinal stenosis. Axial CT photographed at bone window settings demonstrates large osteophytes *(arrows)* derived from the articular processes. The osteophytes are causing severe spinal stenosis with narrowing of the transverse diameter of the canal. The asymmetry of the scanning plane is due to severe scoliosis and has led to asymmetric visualization of the pedicles and vertebral body.

ters measuring 10 to 12 mm are considered to have relative stenosis. This group may develop symptoms when minimal degenerative changes occur. Although measurements may be useful, it is often helpful to compare the size of the spinal canal to that of the thecal sac since a small bony canal may not cause symptoms when the thecal sac is also small.

Degenerative central stenosis may be caused by soft-tissue encroachment upon the thecal sac (thickened ligamentum flavum, bulging disc) as well as bony compression (hypertrophy of articular processes). A patient with central stenosis due to ligamentous hypertrophy may have a normal AP diameter of the osseous canal. Therefore, measurement of this diameter is not reliable in determining the presence of degenerative stenosis. Only 20% of patients with degenerative central stenosis have a decreased AP diameter of the spinal canal (less than 13 mm) as measured by CT.[2] Measurement of cross-sectional area of the thecal sac is a more reliable method of evaluating stenosis. In one study the cross-sectional area of the thecal sac was determined by CT, and the following conclusions were drawn:[2] (1) if the area is 100 mm^2 or less, central lumbar stenosis is present; (2) if the area measures between 100 and 130 mm^2, it is likely that there is early stenosis; (3) an area of 180 ± 50 mm^2 is considered normal. These data highlight the importance of studying the size of the thecal sac; however, the diagnosis of central stenosis can be made without these absolute measurements. The CT demonstration of thickened ligamentum flavum, bulging disc, and/or hypertrophied articular processes compressing the thecal sac is sufficient to make the diagnosis of degenerative spinal stenosis. In evaluating patients who have not had previous surgery, some authors have noted the absence of epidural fat as another sign of stenosis signifying a decrease in effective space.[4,8] Myelography is another reliable method of diagnosing central spinal stenosis. When a myelogram is performed initially and demonstrates a block, CTM can provide important additional information regarding the etiology of the block (Figs. 15F, 15G).

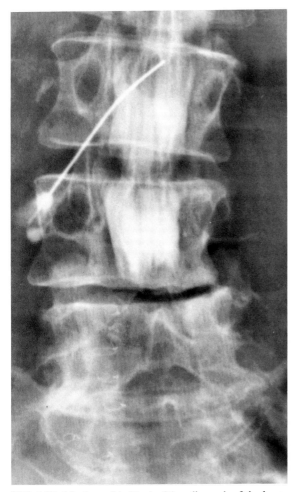

FIG. 15F. Myelographic block. PA radiograph of the lower lumbar spine as part of a myelographic study performed with water-soluble contrast. Contrast was introduced into the subarachnoid space at the L2-L3 level. Complete myelographic block is present at the L4-L5 intervertebral disc level.

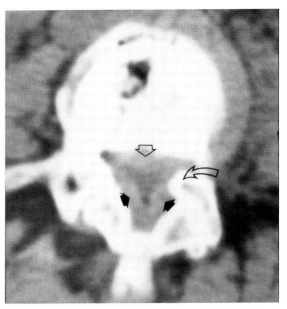

FIG. 15G. Degenerative central spinal stenosis. Same patient as in Fig. 15F. CTM followed myelogram. This scan is through the plane of the myelographic block and demonstrates prominent ligamenta flava *(closed arrows)*, bulging of the disc *(open straight arrow)*, and hypertrophy of the superior articular process of L5 on the left *(curved arrow)*. This combination of findings has led to severe central stenosis and is the cause of the myelographic block. This type of graphic information assists the surgeon in preoperative planning.

Most authors attribute the symptoms of spinal stenosis to pressure on nerve roots within the cauda equina or outside the dura. Other authors suggest the possibility of vascular as well as neural entrapment within the spinal canal or nerve root canal as a cause of symptoms.[5] The classic symptoms of spinal stenosis include back pain with either claudication or sciatic pain in the legs.[13] The pain is typically present when the patient stands or walks and is relieved by lying down or sitting. Activities that require flexion of the torso such as walking uphill or riding a bike are performed without symptoms, whereas hyperextension leads to severe pain. Compared to patients with disc herniation, patients with spinal stenosis tend to have chronic back pain prior to development of radiating leg pain. Symptoms are more frequently bilateral, and reduced straight leg raising is a less frequent finding.[9]

References

1. Arnoldi CC, Brodsky AE, Cauchoix J, et al: Lumbar spinal stenosis and nerve root entrapment syndromes: Definition and classification. *Clin Orthop* 1976;115:4–5.
2. Bolender NF, Schönström NSR, Spengler DM: Role of computed tomography and myelography in the diagnosis of central spinal stenosis. *J Bone Joint Surg Am* 1985;67-A:240–246.
3. Eubanks BA, Cann CE, Brant-Zawadzki M: CT measurement of the diameter of spinal and other bony canals: Effects of section angle and thickness. *Radiology* 1985;157:243–246.
4. Helms CA, Vogler JB: Computed tomography of spinal stenoses and arthroses. *Clin Rheum Dis* 1983;9:417–441.
5. Kirkaldy-Willis WH, McIvor GWD: Editorial comment: Lumbar spinal stenosis. *Clin Orthop* 1976;115:2–3.
6. Kirkaldy-Willis WH, Paine KWE, Cauchoix J, et al: Lumbar spinal stenosis. *Clin Orthop* 1974;99:30–50.
7. Mall JC, Kaiser JA, Heithoff KB: Postoperative spine, in Newton TH, Potts DG (eds): *Computed Tomography of the Spine and Spinal Cord*. San Anselmo, Calif, Clavadel Press, 1983, pp 187–204.
8. McAfee PC, Ullrich CG, Yuan HA, et al: Computed tomography in degenerative spinal stenosis. *Clin Orthop* 1981;161:221–234.
9. Paine KWE: Clinical features of lumbar spinal stenosis. *Clin Orthop* 1976;115:77–82.
10. Postacchini F, Pezzeri G, Montanaro A, et al: Computerized tomography in lumbar stenosis: A preliminary report. *J Bone Joint Surg Br* 1980;62-B:78–82.
11. Rosenbloom S, Cohen WA, Marshall C, et al: Imaging factors influencing spine and cord measurements by CT: A phantom study. *AJNR* 1983;4:646–649.
12. Verbiest H: The significance and principles of computerized axial tomography in idiopathic developmental stenosis of the bony lumbar vertebral canal. *Spine* 1979; 4:369–378.
13. Wiltse LL, Kirkaldy-Willis WH, McIvor GWD: The treatment of spinal stenosis. *Clin Orthop* 1976;115:83–91.

CASE 16

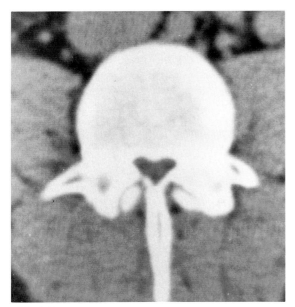

FIG. 16A. Axial CT at L4.

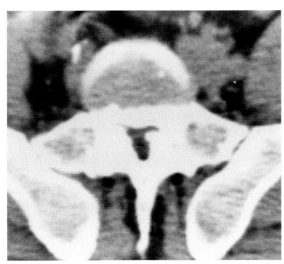

FIG. 16B. Axial CT at L5-S1.

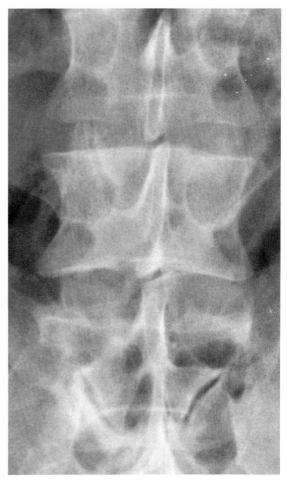

FIG. 16C. AP radiograph of the lower lumbar spine.

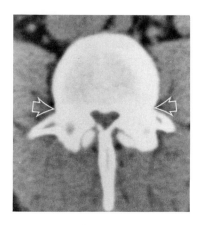
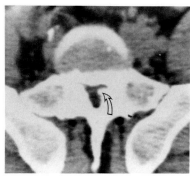
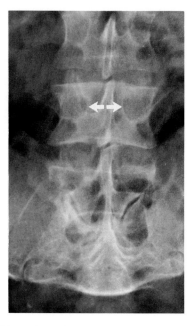

FIG. 16A-1. Achondroplasia with spinal stenosis. Spinal stenosis is evident and is typical of the congenital-developmental type. The spinal canal is small, especially in the anteroposterior dimension. Short broad pedicles *(arrows)* are typical of this type of stenosis.

FIG. 16B-1. Combined developmental and degenerative stenosis. In addition to the findings of developmental stenosis, there are degenerative changes making this a combined form of spinal stenosis. A large osteophyte *(arrow)* derived from an articular process on the left is causing additional stenosis.

FIG. 16C-1. The interpedicular distance is diminished *(arrows)* and decreases progressively at the more caudal levels. This abnormality is typical of achondroplasia. Normally the interpedicular distance increases as successive lower lumbar levels are measured.

Achondroplasia

There are findings typical of congenital spinal stenosis present throughout the lumbar spine in this patient with achondroplasia (Fig. 16A-1). Degenerative changes are also present, producing additional significant narrowing of the canal, making this a combined form of congenital and degenerative stenosis (Fig. 16B-1). Achondroplasia is a classic example of congenital spinal stenosis. It is a congenital disorder of endochondral bone formation that leads to the development of short thick pedicles, which is more pronounced in the lower lumbar spine.[1,2,4] This along with early fusion of the neurocentral synchondroses causes spinal stenosis, which is most severe in the AP dimension. The interpedicular distance is also diminished and narrows progressively from L1 to L5 (Fig. 16C-1), the opposite of the situation in normal individuals. The intrinsically normal spinal cord and nerve roots are contained within a small bony canal.

The thoracolumbar spine is the most frequently symptomatic region in patients with achondroplasia.[4] Neurologic symptoms referable to the thoracolumbar spine are uncommon in childhood or early adulthood and usually occur after age 40 years when degenerative changes compromise the already limited effective space of the spinal and neural canals.[1,2] CT examination of the thoracolumbar spine in patients with achondroplasia demonstrates central and lateral spinal stenosis. These patients have short thick pedicles, thick laminae, hypertrophy of articular processes, marginal vertebral body osteophytes, diminished epidural fat, and a very small subarachnoid space.[3,5] Bulging of the disc contributes to the stenosis and may be seen at multiple levels, frequently in the upper and mid-lumbar region.[5] Scoliosis, thoracolumbar kyphosis, lumbar lordosis, and severe gibbus deformity may be present.[1]

The next most frequent sites of spinal involvement in achondroplasia are the foramen magnum and the cervical spine, areas of particular importance in neonates and infants. CT findings in this area include the presence of a very small foramen magnum and basilar impression.[4,5] Degenerative changes of the lower cervical spine and atrophy of the spinal cord may be demonstrated.[6]

When symptoms referable to spinal stenosis develop in patients with achondroplasia, early surgical intervention is important since delay in treatment may lead to irreversible neurologic damage.[1,2] Presurgical planning must include evaluation of the critical areas of stenosis. Myelography can be used for this purpose; however, a C1-C2 approach may be needed since a lumbar puncture is very difficult and often unsuccessful in patients with achondroplasia because of the severe stenosis.[4,6] When myelography is unsuccessful or technically unsatisfactory, CT has been used to evaluate fully the extent of spinal stenosis and has accurately directed surgical decompression in patients with achondroplasia.[4]

References

1. Alexander E Jr.: Significance of the small lumbar spinal canal: Cauda equina compression syndromes due to spondylosis: Achondroplasia. *J Neurosurg* 1969;31:513–519.
2. Grabias S: The treatment of spinal stenosis. *J Bone Joint Surg Am* 1980;62-A;308–313.
3. Lutter LD, Lonstein JE, Winter RB, et al: Anatomy of the achondroplastic lumbar canal. *Clin Orthop* 1977;126:139–142.
4. Morgan DF, Young RF: Spinal neurological complications of achondroplasia: Results of surgical treatment. *J Neurosurg* 1980;52:463–472.
5. Naidich TP, McLone DG, Harwood-Nash DC: Systemic malformations, in Newton TH, Potts DG (eds): *Computed Tomography of the Spine and Spinal Cord*. San Anselmo, Calif, Clavadel Press, 1983, pp 367–381.
6. Suss RA, Udvarhelyi GB, Wang H, et al: Myelography in achondroplasia: Value of a lateral C1-2 puncture and non-ionic, water-soluble contrast medium. *Radiology* 1983;149:159–163.

CASE 17

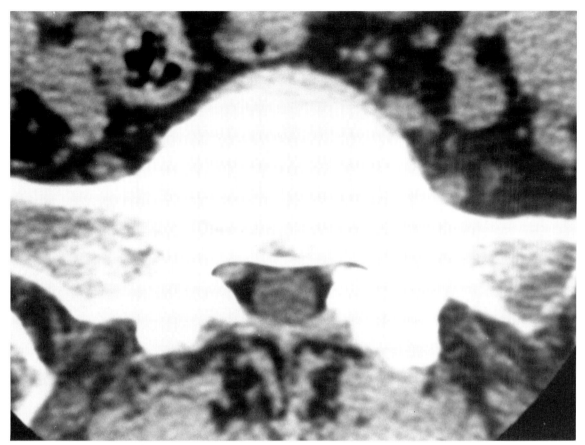

FIG. 17A. Axial CT caudad to the L5-S1 disc level. This patient had left hip and buttock pain.

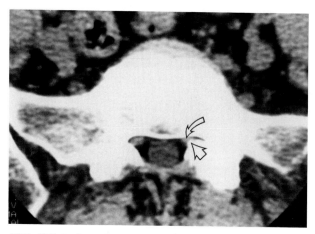

FIG. 17A-1. Lateral recess stenosis. Hypertrophy of the superior articular process of S1 on the left *(straight arrow)* has caused severe lateral recess stenosis with compression of the left S1 nerve root *(curved arrow)*. Compare with the opposite side where the uncompromised nerve root is seen within a normal lateral recess.

Lateral Recess Stenosis

There is severe lateral recess stenosis with nerve root compression due to hypertrophy of the superior articular process (Fig. 17A-1). As the nerve root courses obliquely in a caudal and lateral direction from the thecal sac to the neural foramen, it traverses an area termed the lateral recess, which is bordered anteriorly by the posterolateral margin of the vertebral body and disc, posteriorly by the superior articular process, and laterally by the pedicle[2,8] (Figs. 17B–17D). The L5 nerve root, for example, lies in the lateral recess below the L4-L5 disc space and is bordered laterally by the L5 pedicle and posteriorly by the superior articular process of L5. The narrowest part of the lateral recess is at the superior rostral border of the corresponding pedicle, and the nerve root is most readily compromised by medial hypertrophy of the superior articular process in this region[2,8] (Fig. 17E). Bony or soft-tissue encroachment of the lateral recess leads to lateral recess stenosis, which can best be evaluated by CT.[1,7,10]

Lateral recess stenosis is caused by hypertrophic changes of the superior articular process, marginal vertebral body osteophytes, and/or subluxation of the facet joint[1,5,7] (Fig. 17F). Bulging of the disc, which usually does not cause nerve root compromise when present as an isolated finding, is frequently seen in combination with hypertrophic bony changes of the lateral recess and does lead to further nerve root compression in this setting.[4] The lateral recess can be measured on the transaxial CT examination. A CT measurement of 5 mm or more is normal, 4 mm is borderline, and less than 3 mm is definitely narrow.[5] A trefoil-shaped canal is a common nonpathologic

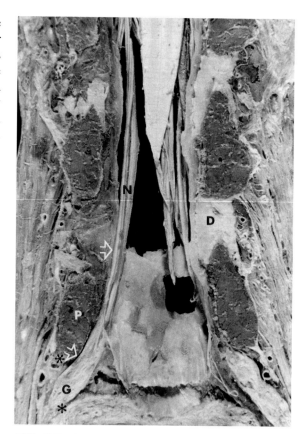

FIG. 17B. Anatomic specimen of the lower lumbar spine is shown in coronal section demonstrating the course of nerve roots within the lateral recess. The laminae and articular processes have been removed, and the specimen is viewed from behind. *D*, intervertebral disc; *G*, dorsal root ganglion; *N*, nerve root; *P*, pedicle; *arrows*, approximate superior and inferior boundaries of the lateral recess; *asterisks*, superior and inferior boundaries of neural foramen.

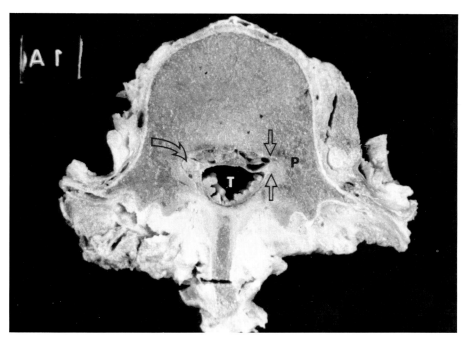

FIG. 17C. Gross anatomic specimen of a lumbar vertebra in the axial plane through the level of the lateral recess pedicle; *T*, thecal sac; *straight arrows,* laterial recess; *curved* lateral recess P, pedicle *curved arrow,* nerve root.

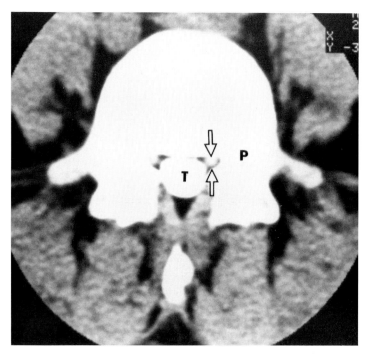

FIG. 17D. Normal lateral recess. Axial CTM at the L5 vertebral level. Contrast fills the thecal sac and is present in the L5 nerve root sleeves. At this level, the nerve roots are within the lateral recess. The height of the lateral recess is normally greater than 3 mm. *P*, pedicle; *T*; thecal sac; *arrows,* height of the lateral recess.

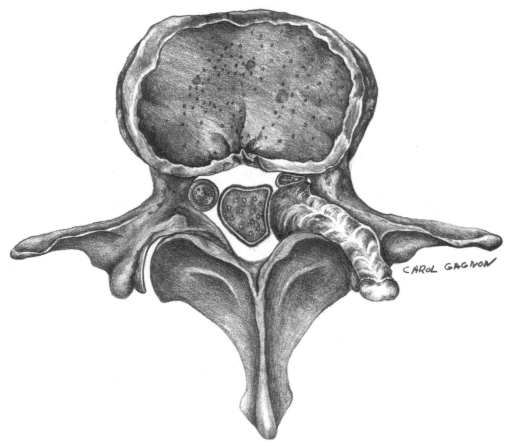

FIG. 17E. Lateral recess and central spinal stenosis. Hypertrophy of the superior articular process causes narrowing of the lateral recess and compression of the nerve root within the recess. These hypertrophic changes may also compress the thecal sac, causing central stenosis.

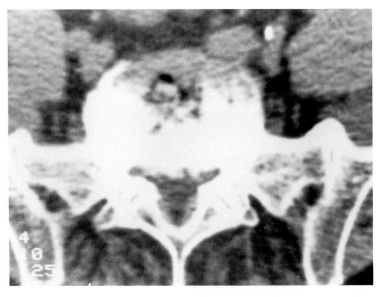

FIG. 17F. Lateral recess stenosis. Bilateral stenosis of the lateral recess at L5-S1 due to osteophytes derived from the posterior superior aspect of the sacrum and hypertrophy of the articular processes. Incidentally noted are osteophytes of the anterior sacroiliac joints.

condition seen in approximately 15% of skeletons examined and is especially prevalent at the L5 level.[3] Although the trefoil shape by itself is not a cause of stenosis, it may predispose to acquired lateral recess stenosis with associated osteophytes and bulging disc. Lateral recess stenosis occurs most frequently at L4-L5 and L5-S1.[4] At the L5-S1 level the S1 nerve root can be seen with CT because of the surrounding epidural fat, and thus the effect of stenosis upon the nerve can be visualized directly (Fig. 17A-1). At L4-L5 the L5 nerve root is not usually seen because of a lack of abundant perineural fat. Thus, the effect of hypertrophic changes on the nerve root can only be inferred unless CTM is performed.[4]

CT is more sensitive and accurate than conventional myelography in identifying lateral recess stenosis.[1,7,10] Myelography may reveal flattening of the nerve root and nerve root sheath beneath the medial aspect of the base of the superior articular process; however, this is often misinterpreted as disc herniation.[2] The use of CT dramatically increases the ability to assess more accurately the lateral recess preoperatively as well as to evaluate the results of surgery.[1,10] This is important because patients with disc herniation frequently have coexistent lateral and/or central stenosis.[1,5,6,9] Failure to recognize or adequately treat lateral spinal stenosis is a major cause of the failed back surgery syndrome (FBSS).[1,5,6,9] In one series of patients diagnosed as having disc herniation, more than half of those who developed FBSS had lateral spinal stenosis either alone or in combination with disc herniation.[1] CT studies have shown that poor surgical results are often due to failure to unroof the lateral recess adequately by extending the surgical decompression laterally.[7] A bilateral surgical approach may be warranted in treating lateral recess stenosis that appears to be unilateral since in the postoperative period there is often contralateral involvement.[1]

Clinically, patients with lateral recess stenosis have severe unilateral or bilateral leg pain, which is brought on by standing and walking and relieved by squatting, sitting in a flexed position, and lying down with hips flexed.[2,8] This symptomatology is referred to as neurogenic claudication. Physical examination reveals an absence of significant objective neurologic abnormalities in the majority of patients, although mild motor deficit and mild abnormal straight leg raising may be present.[8]

References

1. Burton CV, Kirkaldy-Willis WH, Yong-Hing K, et al: Causes of failure of surgery on the lumbar spine. *Clin Orthop* 1981;157:191–199.
2. Ciric I, Mikhael MA, Tarkington JA, et al: The lateral recess syndrome: A variant of spinal stenosis. *J Neurosurg* 1980;53:433–443.
3. Eisenstein S: The trefoil configuration of the lumbar vertebral canal: A study of South African skeletal material. *J Bone Joint Surg Br* 1980;62-B:73–77.
4. Heithoff KB: High-resolution computed tomography of the lumbar spine. *Postgrad Med* 1981;70:193–213.
5. Kirkaldy-Willis WH, Wedge JH, Yong-Hing K, et al: Lumbar spinal nerve lateral entrapment. *Clin Orthop* 1982;169:171–178.
6. Macnab I: Negative disc exploration: An analysis of the causes of nerve-root involvement in sixty-eight patients. *J Bone Joint Surg Am* 1971;53-A:891–903.
7. McAfee PC, Ullrich CG, Yuan HA, et al: Computed tomography in degenerative spinal stenosis. *Clin Orthop* 1981;161:221–234.
8. Mikhael MA, Ciric I, Tarkington JA, et al: Neuroradiological evaluation of lateral recess syndrome. *Radiology* 1981;140:97–107.
9. Quencer RM, Murtagh FR, Post MJD, et al: Postoperative bony stenosis of the lumbar spinal canal: Evaluation of 164 symptomatic patients with axial radiography. *AJR* 1978;131:1059–1064.
10. Risius B, Modic MT, Hardy RW Jr, et al: Sector computed tomographic spine scanning in the diagnosis of lumbar nerve root entrapment. *Radiology* 1982;143:109–114.

CASE 18

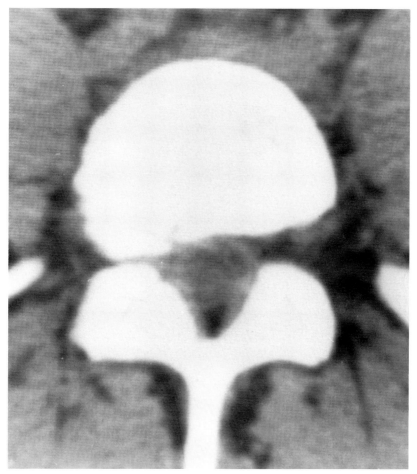

FIG. 18A. CT scan 6 mm above the L5-S1 intervertebral disc space. This 40-year-old man had a 9-month history of severe right leg pain.

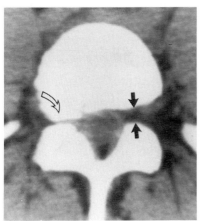

FIG. 18A-1. Stenosis of the neural foramen. There is a large osteophyte on the right *(curved arrow)* derived from the inferior aspect of the L5 vertebral body causing encroachment on the neural foramen. Note the normal appearance of the opposite neural canal *(straight arrows)*.

Stenosis of the Neural Foramen

There is a large osteophyte derived from the inferior aspect of the L5 vertebral body on the right, causing stenosis of the neural foramen (Fig. 18A-1). Stenosis of the neural foramen (more accurately referred to as the lateral nerve root canal) is best identified by CT examination.[7–9] The neural foramen is bordered superiorly by the pedicle of the vertebra above, inferiorly by the pedicle of the vertebra below, and anteriorly by the posterior aspect of the vertebral bodies and the intervertebral disc[5] (Fig. 18B). The posterior boundary is formed by the pars interarticularis and the apex of the superior articular process of the inferior vertebral body (Figs. 18C, 18D). Any bony or soft-tissue encroachment of the neural foramen leads to foraminal stenosis. However, because the nerve root exits through the upper portion of the foramen (approximately 1 to 1.5 cm cephalad to the intervertebral disc) (Fig. 18B), stenosis involving the inferior portion of the foramen may be asymptomatic.

The process that results in osseous stenosis of the neural foramen may begin with degenerative changes leading to degradation of the nucleus pulposus and annulus fibrosus with resultant loss of disc height.[2,6] This causes additional stress on the facet joints, eventually leading to degenerative arthritis and subsequent subluxation of these joints. The superior articular process then moves cephalad and anteriorly, causing narrowing of the neural foramen.[2,6]

Conventional radiography and myelography are inadequate for evaluation of lateral nerve root canal stenosis of the lumbar spine, whereas CT examination is capable of demonstrating this abnormality.[7–9] The CT study may reveal hypertrophic changes derived

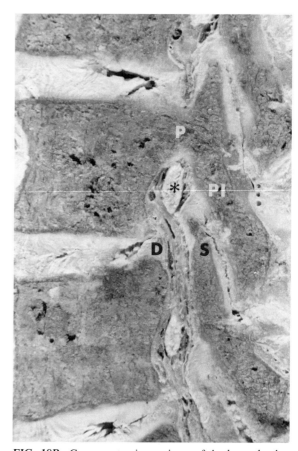

FIG. 18B. Gross anatomic specimen of the lower lumbar vertebrae in the parasagittal plane demonstrating the nerve root *(asterisk)* exiting the superior portion of the neural foramen just beneath the pedicle *(P)*. Note the relationship of the nerve root to the pars interarticularis *(PI)* of the same vertebral level, the superior articular process of the vertebra below *(S)*, and the intervertebral disc *(D)*.

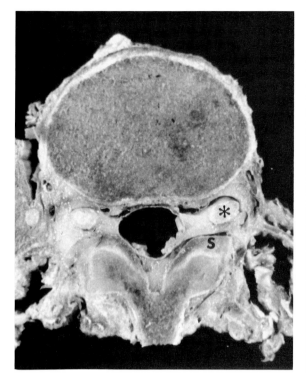

FIG. 18C. Gross anatomic specimen of the L4 vertebra in the axial plane through the level of the neural foramen. The dorsal root ganglion *(asterisk)* is surrounded by fat and has a close relationship to the superior articular process of L5 *(S)*.

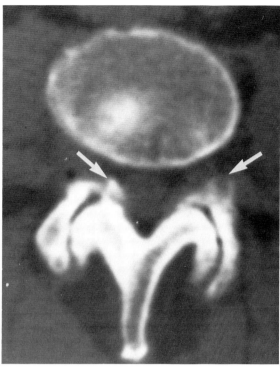

FIG. 18E. Stenosis of the neural foramen and degenerative arthritis of facets in a patient with degenerative spondylolisthesis at L4-L5. CT cephalad to the L4-L5 disc at the level of the neural foramen. There are bilateral hypertrophic changes of the superior articular processes of L5 *(arrows)* and narrowing of the facet joints. Note the somewhat sagittal orientation of the facet joints. The degree of neural canal stenosis was better evaluated by the reconstruction technique shown in Fig. 18F.

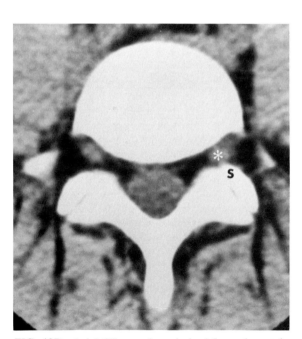

FIG. 18D. Axial CT scan through the L5 vertebra at the level of the neural foramen. The dorsal root ganglion *(asterisk)* exits the neural foramen bordered anteriorly by the posterior margin of the vertebral body and posteriorly by the superior articular process of S1 *(S)*. Fat surrounds the dorsal root ganglion.

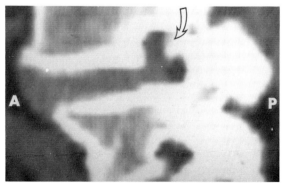

FIG. 18F. Stenosis of the neural foramen. Same patient as in Fig. 18E evaluated with oblique reconstruction of axial images through the plane of the left neural foramen. Encroachment upon the superior portion of the neural foramen is caused by osseous hypertrophy derived from the superior articular process of L5 *(arrow)*. *A*, anterior; *P*, posterior.

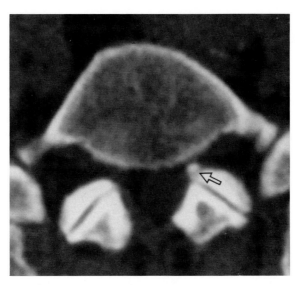

FIG. 18G. Stenosis of the neural foramen. This patient had previous surgery for central spinal stenosis; however, decompression did not extend to the neural canal. Hypertrophy of the superior articular process (arrow) is causing narrowing of the left neural canal, which was subsequently unroofed at reoperation.

from the superior articular process, or a hard disc representing either posterior vertebral body osteophyte or calcified disc material (Fig. 18E). The dorsal nerve root ganglion may appear flattened or enlarged, and the perineural fat within the foramen may be obliterated.[3] The grading of stenosis involving the lateral nerve root canal is based on the amount of narrowing of the canal and the degree of obliteration of perineural fat.[9] One word of caution: Actual measurements of neural foraminal width from axial CT scans are influenced by several factors (e.g., CT window settings and reference points for measurement), which may lead to interobserver discrepancies.[1] Sagittal and oblique reformation of axial images may lead to increased diagnostic accuracy of osseous stenosis of the neural foramen[3] (Fig. 18F). Other causes of foraminal nerve root entrapment may similarly be associated with radiculopathy and a negative myelogram and include lateral disc herniation, postoperative fibrotic scarring, spondylolisthesis, and tumor.[4,8]

As was the case with lateral recess stenosis (Case 17), patients with the failed back surgery syndrome (FBSS) may have stenosis of the neural foramen that was either inaccurately diagnosed preoperatively or inadequately decompressed (Fig. 18G). It is not certain whether the lateral stenosis is present prior to surgery or is the result of surgery in patients with FBSS. Some authors, however, believe that in most cases the stenosis precedes the surgery.[3] With the advent of CT and the increased ability to diagnose lateral stenosis preoperatively, a decrease in the incidence of the FBSS has been noted by some.[3]

References

1. Beers GJ, Carter AP, Leiter BE, et al: Interobserver discrepancies in distance measurements from lumbar spine CT scans. *AJR* 1985;144:395–398.
2. Burton CV, Kirkaldy-Willis WH, Yong-Hing K, et al: Causes of failure of surgery on the lumbar spine. *Clin Orthop* 1981; 157:191–199.
3. Heithoff KB: High-resolution computed tomography and stenosis: An evaluation of the causes and cures of the failed back surgery syndrome, in Post MJD (ed): *Computed Tomography of the Spine*. Baltimore, Williams & Wilkins, 1984, pp 506–545.
4. Helms CA, Vogler JB: Computed tomography of spinal stenoses and arthroses. *Clin Rheum Dis* 1983;9:417–441.
5. Kirkaldy-Willis WH, McIvor GWD: Editorial comment: Lumbar spinal stenosis. *Clin Orthop* 1976; 115:2–3.
6. Kirkaldy-Willis WH, Wedge JH, Yong-Hing K, et al: Lumbar spinal nerve lateral entrapment. *Clin Orthop* 1982;169:171–178.
7. McAfee PC, Ullrich CG, Yuan HA, et al: Computed tomography in degenerative spinal stenosis. *Clin Orthop* 1981;161: 221–234.
8. Osborne DR, Heinz ER, Bullard D, et al: Role of computed tomography in the radiological evaluation of painful radiculopathy after negative myelography: Foraminal neural entrapment. *Neurosurgery* 1984;14:147–153.
9. Risius B, Modic MT, Hardy RW Jr, et al: Sector computed tomographic spine scanning in the diagnosis of lumbar nerve root entrapment. *Radiology* 1982;143:109–114.

CASE 19

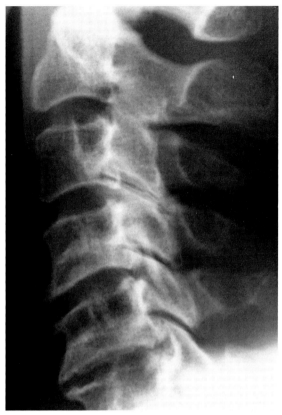

FIG. 19A. Lateral radiograph of the cervical spine in a 64-year-old Oriental male who presented with spastic gait and a 3-month history of bilateral upper and lower extremity dysesthesias.

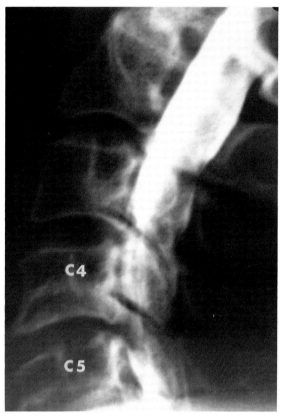

FIG. 19B. Cross-table lateral radiograph of the cervical spine from a myelographic study.

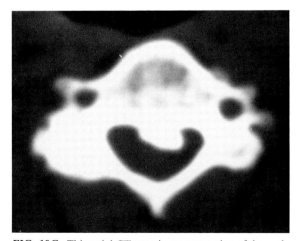

FIG. 19C. This axial CT scan is representative of the multiple scan sections obtained throughout the C4 and C5 levels.

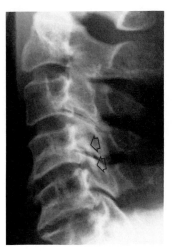
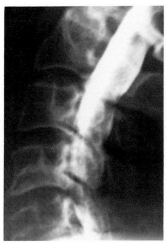
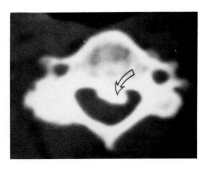

FIG. 19A-1. Ossification of the posterior longitudinal ligament (OPLL). Careful examination is needed to detect the longitudinal ossification *(arrows)* posterior to the C4 and C5 vertebral bodies.

FIG. 19B-1. Partial myelographic block is demonstrated at C4 and C5.

FIG. 19C-1. OPLL is well demonstrated by CT. The ossified ligament *(arrow)* is causing spinal stenosis. This scan is representative of the additional scans obtained at the C4 and C5 levels, all showing OPLL.

Ossification of the Posterior Longitudinal Ligament

In addition to degenerative changes of the cervical spine, there is a longitudinal calcification or ossification seen posterior to the C4 and C5 vertebral bodies on the lateral radiograph (Fig. 19A-1). Partial myelographic block is present at C4 and C5 (Fig. 19B-1). A representative axial CT scan at the C4 level demonstrates a large ossified mass in the midline adjacent to the posterior aspect of the vertebral body (Fig. 19C-1). This type of ossification was found in all axial scans from the superior aspect of C4 to the inferior aspect of C5 and represents ossification of the posterior longitudinal ligament (OPLL). Marked stenosis of the spinal canal and compression of the spinal cord are demonstrated and are the cause of myelopathy in this patient, who presented with spastic gait.

OPLL is a recently recognized clinical disorder that is more prevalent in Japan and the eastern Asian countries, where it occurs in 1% to 3% of the population with cervical symptoms.[5,7] The incidence among Caucasians in only 0.2%.[5,7] The cervical spine is much more frequently involved than the thoracic or lumbar spine. Within the cervical spine, OPLL may occur at any level; however, the most frequent site of ossification is C5, followed by C4, C6, and C3.[7] Usually more than one cervical level is involved. OPLL causes compression of the spinal cord leading to myelopathy and radiculopathy. Compression of the anterior spinal arteries may also be a cause of symptoms.[3] The most common symptoms at presentation are numbness of the hands, weakness of the legs, difficulty walking (spastic gait), pain in the neck and arms, and urinary or intestinal symptoms.[1,3] Some patients are asymptomatic.

The lateral radiograph of the spine is evaluated for the presence of a bony density having a longitudinal orientation posterior to the vertebral bodies. Several types of ossification have been described: segmental (behind each vertebral body), continuous (involving multiple vertebrae without interruption), mixed segmental and continuous, and localized (limited to the intervertebral disc level).[7] The localized form is the least frequent type and may be difficult to distinguish from spondylosis. CT is utilized to determine the configuration of the ossified mass, the degree of canal compromise, and the longitudinal extent of ossification. CT may be particularly helpful in the upper thoracic spine where ossification is difficult to detect with conventional radiography. Because OPLL occurs much less frequently then degenerative spondylosis, the latter diagnosis may be made incorrectly in patients with OPLL evaluated by conventional radiography and myelography. Typically, however, the CT examination demonstrates that the ossified mass extends posterior to the entire longitudinal extent of

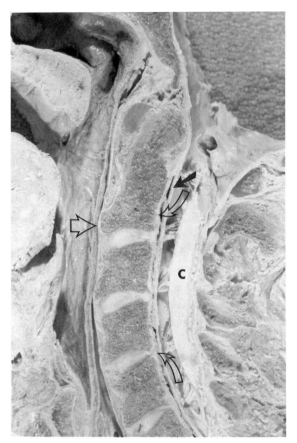

FIG. 19D. Anatomic specimen of the upper cervical spine in the near-sagittal plane demonstrating the close relationship of the posterior *(curved arrows)* and anterior longitudinal *(open straight arrow)* ligaments to the vertebrae and discs. C, spinal cord; *closed arrow*, thecal sac.

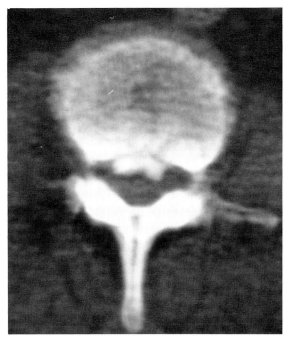

FIG. 19E. OPLL of the lumbar spine. Axial CT at the vertebral endplate level. The ossified ligament appears attached to the posterior vertebral body.

the vertebral body rather than being limited to the disc level as is typically seen with degenerative spondylosis.

Anatomically, the posterior longitudinal ligament extends from the skull to the sacrum along the posterior aspect of the intervertebral discs and vertebral bodies (Fig. 19D). It is present in the midline and attaches to the posterior aspect of the discs and vertebral body margins, being separated from the midportion of the vertebral bodies by retrovertebral venous structures. OPLL most frequently occurs in the midline and typically has an ovoid or oblong shape on axial CT.[5] The ossification may appear attached to the vetebral body at some levels and unattached at others[3,5] (Figs. 19E, 19F). The gap between the vertebral body and the ossified ligament is thought to represent an unossified deep layer of the posterior longitudinal ligament[3,6] or may be due to interposed venous structures. Some authors have described a tandem type of ossification with two layers of ossifica-

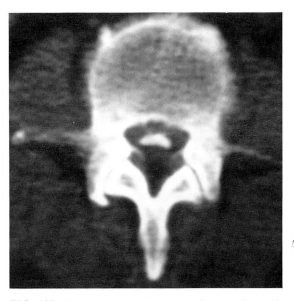

FIG. 19F. OPLL of the lumbar spine. Same patient as in Fig. 19E. Axial CT at the midvertebral level. The ossified ligament is obviously separate from the posterior vertebral body. The unossified area between the vertebral body and the ossified ligament may be caused by an unossified deep layer of the ligament or by interposed venous structures.

Case 19

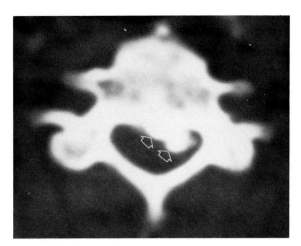

FIG. 19G. OPLL. Same patient as in Fig. 19A-1. Cervical scan demonstrates the tandem type of OPLL with two layers of ossification demonstrated *(arrows)*.

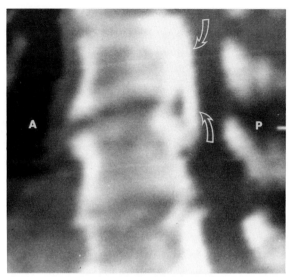

FIG. 19H. OPLL. Same patient as in Fig. 19A-1. Sagittal reconstruction of the axial images can be used to delineate further the longitudinal extent of the ligamentous ossification *(arrows)* as well as the degree of spinal stenosis. A, anterior; P, posterior.

tion seen on axial CT[5] (Fig. 19G). This may represent the mixed continuous and segmental type of ossification, with a continuous layer of ossification extending posterior to a separate segmental layer of ossification.[5]

Correlation between the degree of spinal canal compromise and clinical symptoms suggests that severe myelopathy is most likely to occur when the axial stenosis exceeds 30%.[1] Sagittal reconstruction of the axial images can further define the longitudinal extent of the ossification as well as the degree of spinal canal compromise (Fig. 19H). Some patients, however, have a paucity of symptoms despite severe canal stenosis. Treatment may be conservative or surgical, depending on the clinical state. If surgery is contemplated it becomes important to distinguish OPLL from degenerative spondylosis preoperatively since the surgical approaches differ. OPLL can be surgically treated by a posterior or an anterior approach. The anterior approach involves partial resection of the vertebrae, release or removal of the ossified mass, and insertion of an iliac bone graft[2] (Fig. 19I).

The etiology of OPLL is not known, but certain common characteristics of patients with OPLL have been found. These patients tend to have abnormal glucose and calcium metabolism and a generalized hyperostotic state with ossification of other paraspinal ligaments.[7] Ossification or calcification of the posterior longitudinal ligament has also been seen in 50% of patients with diffuse idiopathic skeletal hyperostosis.[6] Ossification of the ligamentum flavum may occur separately or in association with OPLL. Ossification of the ligamentum flavum is most commonly found in the thoracic spine and may also be a cause

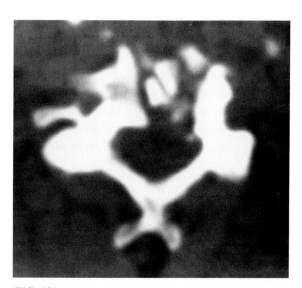

FIG. 19I. Postoperative status of patient treated for OPLL. Same patient as in Fig. 19A-1. There has been partial resection of the vertebral body, removal of the ossified mass, and insertion of an iliac bone graft.

of myelopathy with compression of the spinal cord and nerve roots.[4]

References

1. Hanai K, Adachi H, Ogasawara H: Axial transverse tomography of the cervical spine narrowed by ossification of the posterior longitudinal ligament. *J Bone Joint Surg Am* 1977;59-A:481–484.
2. Hanai K, Inouye Y, Kawai K, et al: Anterior decompression for myelopathy resulting from ossification of the posterior longitudinal ligament. *J Bone Joint Surg Br* 1982;64-B:561–564.
3. Hanna M, Watt I: Posterior longitudinal ligament calcification of the cervical spine. *Br J Radiol* 1979;52:901–905.
4. Hukuda S, Mochizuki T, Ogata M, et al: The pattern of spinal and extraspinal hyperostosis in patients with ossification of the posterior longitudinal ligament and the ligamentum flavum causing myelopathy. *Skeletal Radiol* 1983;10:79–85.
5. Murakami J, Russell WJ, Hayabuchi N, et al: Computed tomography of posterior longitudinal ligament ossification: Its appearance and diagnostic value with special reference to thoracic lesions. *J Comput Assist Tomogr* 1982;6:41–50.
6. Resnick D, Guerra J Jr, Robinson CA, et al: Association of diffuse idiopathic skeletal hyperostosis (DISH) and calcification and ossification of the posterior longitudinal ligament. *AJR* 1978;131:1049–1053.
7. Tsuyama N: Ossification of the posterior longitudinal ligament of the spine. *Clin Orthop* 1984;184:71–84.

CASE 20

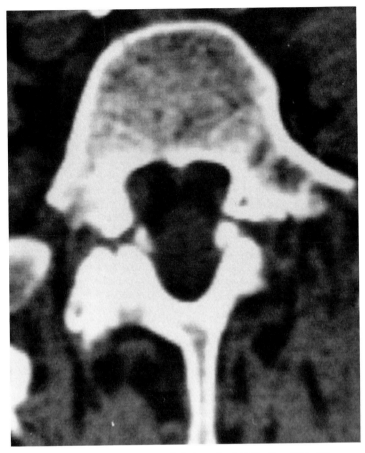

FIG. 20A. Axial CT 10 mm cephalad to the L5-S1 disc. This 57-year-old had symptoms of intermittent neurogenic claudication.

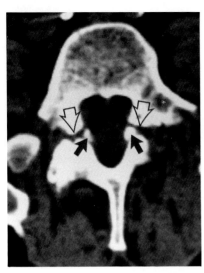

FIG. 20A-1. There is bilateral spondylolysis *(open arrows)*, which is differentiated from facet joints by the jagged, irregular appearance and coronal plane. Osseous fragments adjacent to the medial aspect of the lysis *(closed arrows)* are causing narrowing of the transverse diameter of the canal. These fragments were found to be free at the time of surgery. Note the elongated AP diameter of the canal. This patient had a grade II spondylolisthesis.

Spondylolysis

There is bilateral spondylolysis of L5 (Fig. 20A-1). Spondylolysis is a break in the pars interarticularis. This can be reliably diagnosed by CT even in the absence of confirmation by conventional radiography.[4,6] The CT examination is best viewed at bone window settings (1,000 to 2,000 HU). With transaxial scans spondylolysis may have an appearance that resembles the facet joint; however, several differentiating features have been described.[4] The pars defect of spondylolysis is located 10 to 15 mm above the disc level and has jagged, irregular, noncorticated margins. The facet joint, on the other hand, is located at and adjacent to the disc level and has smooth, straight or slightly curved cortical margins. The pars defect is located anterior to the facet joint and may be in a more coronal plane than the joint (Figs. 20B–20D). The presence of a complete intact cortical ring outlining the bony spinal canal at the level of the inferior aspect of the pedicle excludes the diagnosis of spondylolysis.[6]

Additional CT findings of spondylolysis have been described.[4,8] Callus or granulation tissue adjacent to the pars defect is found in 20% of cases and may cause compression of the thecal sac or nerve root[8] (Figs. 20E, 20F). Laminal fragmentation has been described in approximately 15% of patients with spondylolysis studied by conventional radiography[1] (Fig. 20F). An increased AP diameter of the spinal canal is found with spondylolysis, usually but not always in association with spondylolisthesis. The pars may be narrow or sclerotic. Unilateral spondylolysis may have associated contralateral neural arch sclerosis and hypertrophy (Figs. 20G, 20H). Further delineation of spondylolysis can be obtained by sagittal reconstruction (Fig. 20I).

Both congenital and acquired theories have been offered for the etiology of spondylolysis; however, the most popular theory is that of repeated minor trauma causing a break in the pars similar to a stress fracture.[8] This may take place in individuals predisposed to spondylolysis because of a congenitally abnormal pars. Spondylolysis occurs in 5% of the population. The most frequent site of lysis is L5 (90%), followed by L4 (10%), with fewer than 1% occurring at L3.[8] Approximately 60% of patients with spondylolysis have associated spondylolisthesis.[8] Spondylolysis is more commonly bilateral, especially when the lysis is associated with spondylolisthesis.

Isthmic spondylolisthesis is an anterior slippage of one vertebra upon the adjacent caudad vertebra, with an associated break in the pars interarticularis of the superior vertebra. This most commonly occurs at L5-S1 and can be identified with conventional radiography. When interpreting a CT scan, one examines the lateral digital radiograph for spondylolisthesis. The axial scan demonstrates a large amount of disc material posterior to the L5 vertebral body (Fig. 20J). This is the typical appearance of spon-

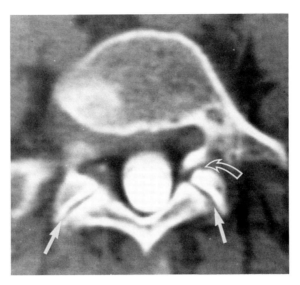

FIG. 20B. Spondylolysis. CTM above the L5-S1 disc level. Slight asymmetry in patient position during scanning accounts for asymmetric visualization of pedicles and disc. The facet joints have an oblique orientation *(straight arrows)*. An additional lucent line due to spondylolysis *(curved arrow)* is seen anterior to the left facet joint. This patient also had grade II spondylolisthesis at L5-S1.

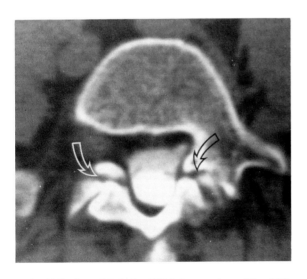

FIG. 20C. Spondylolysis. CTM 4 mm above Fig. 20B. The superior aspect of the facet joints can still be seen. At this level bilateral spondylolysis is noted *(arrows)*. The defects of spondylolysis lie in a more coronal plane than do the facet joints.

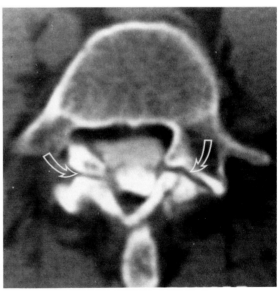

FIG. 20D. Spondylolysis. CTM 8 mm above Fig. 20B. The facet joints are no longer seen. The bilateral linear lucencies are due to spondylolysis *(arrows)*. Examining the sequence of scans permits CT identification of spondylolysis and prevents mistaking the findings for abnormal facet joints.

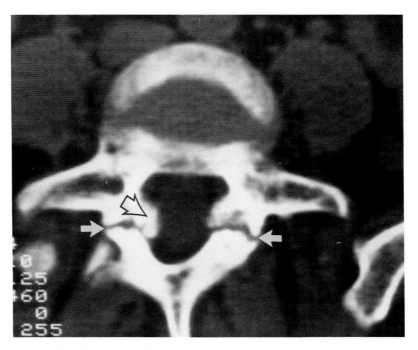

FIG. 20E. Spondylolysis with callus. Axial CT demonstrates bilateral spondylolysis at L5 as irregular, jagged defects having a coronal orientation *(small arrows).* Callus at the site of lysis is more prominent on the right side *(large arrow).*

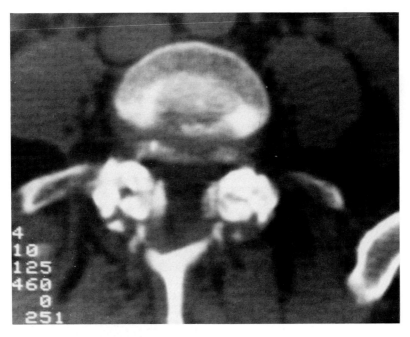

FIG. 20F. Spondylolysis with callus and fragmentation. CT 4 mm cephalad to Fig. 20E. There is osseous encroachment of the spinal canal, and marked fragmentation is identified.

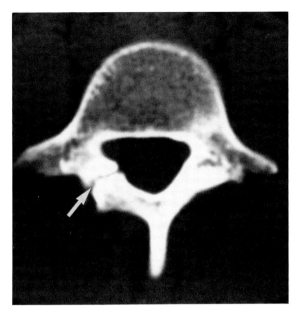

FIG. 20G. Unilateral spondylolysis. Spondylolysis of L5 is present on the right *(arrow)*.

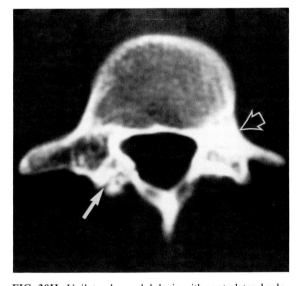

FIG. 20H. Unilateral spondylolysis with contralateral sclerosis and hypertrophy of the pedicle. Same patient as in Fig. 20G. Spondylolysis is again demonstrated on the right *(long arrow)*. This CT is through the plane of the pedicles. Note the hypertrophy and sclerosis of the left pedicle *(short arrow)*, which develop in response to the unilateral spondylolysis.

FIG. 20I. Spondylolysis. Parasagittal reconstruction of axial images through the plane of the defect in the left pars interarticularis. The *vertical white line* in the upper insert identifies the plane of scanning. The reconstructed image reveals a defect in the pars *(arrow)*.

dylolisthesis and should not be confused with disc herniation. More caudad, the axial scan shows the disc sandwiched between the vertebral bodies of L5 (anterior) and S1 (posterior) (Fig. 20K). Typically, the spinal canal appears elongated in the AP dimension.

The role of CT in evaluating patients with isthmic spondylolisthesis includes the diagnosis of associated central and lateral stenosis and the diagnosis of associated disc disease.[3,7] Approximately one third of symptomatic patients with isthmic spondylolisthesis have spinal stenosis, most often neural foraminal or lateral recess stenosis at the same interspace level as the spondylolisthesis.[3] At the L5-S1 level, anterior and caudal displacement of L5 leads to decreased height of the disc and caudal displacement of the L5 pedicle with subsequent compression of the L5 nerve root in the neural canal.[5] Callus and granulation tissue at the site of spondylolysis may encroach on the nerve root at the lateral recess. Central stenosis may occur secondary to thickening of the ligamentum flavum and the laminae at the level of slippage and at the level above the slippage. This may cause compression of the cauda equina.

When the possibility of disc herniation is being considered in patients with isthmic spondylolisthesis, several points should be remembered: The transaxial CT scan will reveal disc material posterior to the superior vertebral body, and this should not be routinely diagnosed as disc herniation. Disc herniation is unusual at the level of spondylolisthesis and occurs much more frequently at the disc level above.[3,8] Although caution should be exercised in diagnosing disc herniation at the level of spondylolisthesis, this diagnosis may be considered if there is asymmetric compression

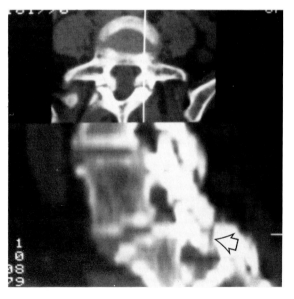

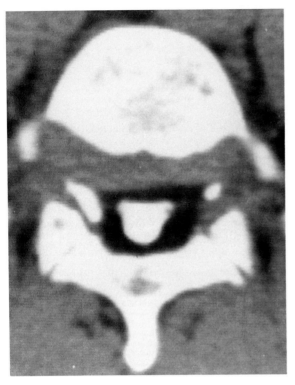 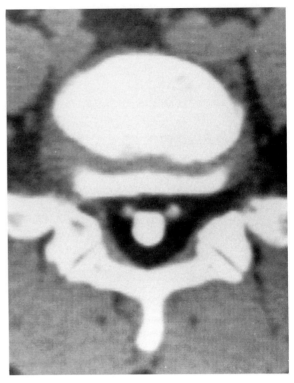

FIG. 20J. Spondylolisthesis with appearance of pseudoherniation of the disc. Axial CTM through the plane of the posterior aspect of the L5-S1 disc. The disc is seen extending far posterior to the L5 vertebral body, as a result of grade II spondylolisthesis and the plane of scanning. This finding does not indicate a disc herniation but rather is a form of pseudoherniation of the disc. Note the presence of epidural fat and the lack of thecal sac compression. Bilateral spondylolysis is also present.

FIG. 20K. Spondylolisthesis with pseudoherniation of the disc. This scan was obtained 4 mm caudad to Fig. 20J. The posterior superior aspect of the sacrum is now visualized in the same relative location as the posterior disc margin seen in Fig. 20J. Note the normal appearance of the S1 nerve roots and the epidural fat. A true disc herniation is not present. In patients with isthmic spondylolisthesis, most disc herniations occur at the disc level above the pars defect.

of the epidural fat, thecal sac, or nerve root.[2] Sagittal reconstruction may be helpful in determining whether the disc extends significantly beyond the posterior aspect of the sacrum.[3,8] Reconstruction views are also helpful in evaluating the previously described stenoses that may accompany isthmic spondylolisthesis.

References

1. Amato M, Totty WG, Gilula LA: Spondylolysis of the lumbar spine: Demonstration of defects and laminal fragmentation. *Radiology* 1984;153:627–629.
2. Braun IF, Lin JP, George AE, et al: Pitfalls in the computed tomographic evaluation of the lumbar spine in disc disease. *Neuroradiology* 1984;26:15–20.
3. Elster AD, Jensen KM: Computed tomography of spondylolisthesis: Patterns of associated pathology. *J Comput Assist Tomogr* 1985;9:867–874.
4. Grogan JP, Hemminghytt S, Williams AL, et al: Spondylolysis studied with computed tomography. *Radiology* 1982;145:737–742.
5. Kirkaldy-Willis WH, Paine KWE, Cauchoix J, et al: Lumbar spinal stenosis. *Clin Orthop* 1974;99:30–50.
6. Langston JW, Gavant ML: "Incomplete ring" sign: A simple method for CT detection of spondylolysis. *J Comput Assist Tomogr* 1985;9:728–729.
7. McAfee PC, Yuan HA: Computed tomography in spondylolisthesis. *Clin Orthop* 1982;166:62–71.
8. Rothman SLG, Glenn WV Jr: CT multiplanar reconstruction in 253 cases of lumbar spondylolysis. *AJNR* 1984;5:81–90.

CASE 21

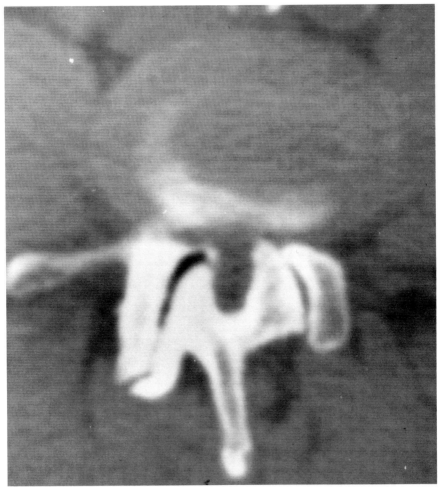

FIG. 21A. CT scan at the L4-L5 intervertebral disc level viewed at bone window settings. This 65-year-old male has chronic low back pain.

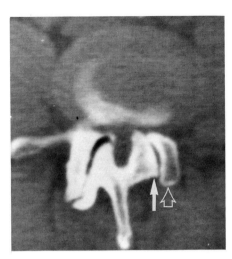

FIG. 21A-1. Degenerative spondylolisthesis with stenosis. The facet joints have a somewhat sagittal orientation, and there is anterior subluxation of the inferior articular process of L4 *(long arrow)* in relation to the superior articular process of L5 *(short arrow)*. Changes of osteoarthritis of the facet joints are present with joint space narrowing, hypertrophy of the articular processes, and vacuum facet phenomenon. The decreased transverse diameter of the spinal canal indicates central spinal stenosis.

Degenerative Spondylolisthesis

The transaxial CT scan demonstrates osteoarthritis of the facet joints and central spinal stenosis (Fig. 21A-1). Note the anterior subluxation of the inferior articular process of L4 in relation to the superior articular process of L5. Sagittal orientation of the facets can also be appreciated. This combination of findings is due to degenerative spondylolisthesis: an anterior slippage of one vertebra on the adjacent caudad vertebra associated with an intact pars interarticularis. This occurs secondary to disc degeneration and facet joint instability, which may be related to congenital malalignment of the facets (e.g., sagittal orientation of the facets).[6] Degenerative changes occur in the facet joints and lead to spinal stenosis.

Degenerative spondylolisthesis most often occurs at L4-L5 (unlike isthmic spondylolisthesis, which is usually at L5-S1). The slippage can be seen on the lateral digital radiograph and is usually limited to a grade I/IV (Fig. 21B). Approximately 30% of symptomatic patients with degenerative spondylolisthesis have significant protrusion of the disc at the same level as the spondylolisthesis[1] (Fig. 21C). Sagittal

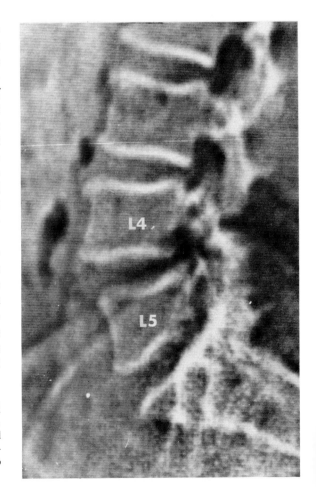

FIG. 21B. Degenerative spondylolisthesis. This lateral digital radiograph reveals a grade I spondylolisthesis at L4-L5. Typical findings of degenerative spondylolisthesis were found on subsequent CT examination. Degenerative spondylolisthesis most often occurs at L4-L5 and is usually limited to grade I severity.

reconstruction may be useful in determining whether the disc protrudes significantly beyond the posterior margin of the inferior vertebral body.[1] CT of patients with degenerative spondylolisthesis also reveals degenerative changes of the facet joints such as narrowing, sclerosis, osteophytes, and vacuum phenomenon. Facet joint subluxation may be present, especially if the facet joints have a sagittal orientation.

The role of CT is degenerative spondylolisthesis includes the evaluation of central and lateral spinal stenosis.[3,4,6] Anterior displacement of the upper vertebra leads to central spinal stenosis with compression of the cauda equina. Lateral stenosis occurs secondary to subluxation of facet joints, osteophyte formation of the articular processes, and bulging of the disc into the nerve root canal (Fig. 21D). Sagittal reconstruction may help in evaluating the severity of the central and lateral stenosis (Fig. 21E).

Degenerative spondylolisthesis usually occurs after age 50 and affects women four times more frequently than men.[5] Patients often have back, buttock, or thigh pain and may have neurogenic claudication or postural symptomatology, the pain occurring with spine extension.[2,4] Approximately 10% of patients with this disorder have severe symptoms that require decompression laminectomy and excision of the medial portion of the articular processes.[5]

References

1. Elster AD, Jensen KM: Computed tomography of spondylolisthesis: Patterns of associated pathology. *J Comput Assist Tomogr* 1985;9:867–874.
2. Epstein BS, Epstein JA, Jones MD: Degenerative spondylolisthesis with an intact neural arch. *Radiol Clin North Am* 1977;15:275–287.
3. McAfee PC, Ullrich CG, Yuan HA, et al: Computed tomography in degenerative spinal stenosis. *Clin Orthop* 1981;161: 221–234.
4. McAfee PC, Yuan HA: Computed tomography in spondylolisthesis. *Clin Orthop* 1982;166:62–71.
5. Rosenberg NJ: Degenerative spondylolisthesis: Predisposing factors. *J Bone Joint Surg Am* 1975;57-A:467–474.
6. Rothman SLG, Glenn WV Jr: Spondylolysis and spondylolisthesis, in Post MJG (ed): *Computed Tomography of the Spine*. Baltimore, Williams & Wilkins, 1984, pp 591–615.

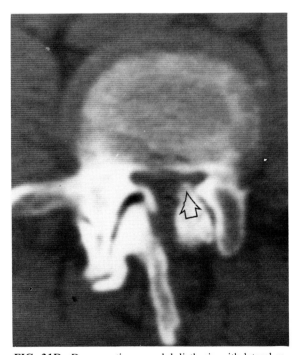

FIG. 21C. Degenerative spondylolisthesis. Axial CT at L4-L5 demonstrates changes of degenerative spondylolisthesis with anterior subluxation of the inferior articular processes of L4 *(I)* in relation to the superior articular processes of L5 *(S)*. There are bilateral hypertrophic changes of the articular processes. There is protrusion of the disc posteriorly causing compression of the thecal sac. Sagittal reconstruction of axial images can be used to differentiate significant disc protrusion from pseudoherniation of the disc.

FIG. 21D. Degenerative spondylolisthesis with lateral recess stenosis. Same patient as in Fig. 21A-1. This scan is caudad to the L4-L5 disc at the level of the pedicles. In addition to central stenosis there is stenosis of the left lateral recess due to anterior subluxation and hypertrophy of the left inferior articular process of L4 *(arrow)*. Osteoarthritis of the facet joints is noted. The incomplete visualization of the left lamina on this scan section is due to slight asymmetry of the scanning plane.

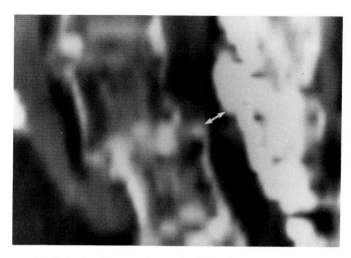

FIG. 21E. Degenerative spondylolisthesis with central stenosis. Sagittal reconstruction of axial images from the same patient seen in Fig. 21C. There is a grade I spondylolisthesis at L4-L5. Central spinal stenosis is apparent with approximately 50% narrowing of the AP diameter of the spinal canal at L4-L5 *(arrows)*. The use of reconstruction technique is important in evaluating the various forms of spinal stenosis.

CASE 22

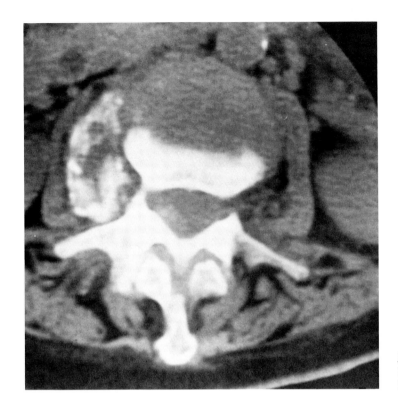

FIG. 22A. Axial CT at the L2-L3 disc level. This 65-year-old woman had had back pain for 18 months.

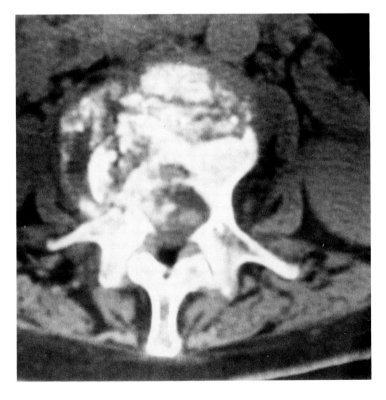

FIG. 22B. Axial CT 5 mm cephalad to Fig. 22A at the L2 level.

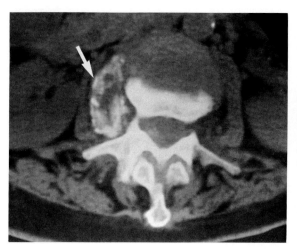 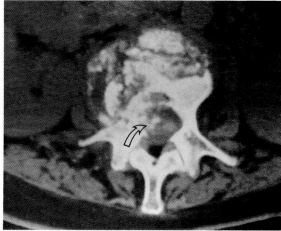

FIG. 22A-1. Tuberculous osteomyelitis. There is a large calcified psoas abscess on the right *(arrow)*.

FIG. 22B-1. There is marked destruction of the L2 vertebral body and the right pedicle. Extensive bony fragmentation is noted with extension into the spinal canal *(arrow)*.

Tuberculous Osteomyelitis

There is vertebral body destruction, bone fragmentation within the spinal canal, and a large, calcified paravertebral abscess (Figs. 22A-1, 22B-1). These findings are due to tuberculosis of the spine. Although the incidence of tuberculosis has decreased in the United States, this diagnosis must still be considered in the evaluation of patients with spinal disorders, especially among those patients who have recently immigrated and those of a low socioeconomic background. Worldwide, tuberculosis of the spine remains quite prevalent. Approximately 1% of patients with tuberculosis have skeletal tuberculosis, and 50% of these patients have spinal involvement.[1] Only 50% of patients with skeletal tuberculosis have pulmonary tuberculosis as well.[1] Tuberculosis of the spine usually has a chronic course, and patients most often present with back pain of several months' to years' duration.[2] Neurologic findings are frequently absent despite widespread disease.[2,5] Some patients, however, have subacute or progressive neurologic symptoms. The lower thoracic spine is most frequently involved, followed by the lumbar spine.[4,5] Hematogenous spread of tuberculosis is probably the primary mode of spinal disease, although direct extension from involved lymph nodes and extension from the subarachnoid space have also been implicated.[5]

The radionuclide bone scan may be useful in localizing the site or sites of infection, but it is normal in 35% of cases.[4] Conventional radiography and CT can demonstrate osteolysis, disc space narrowing, paravertebral soft-tissue mass, multilevel involvement, and kyphosis.[1-5] The early findings are typically either osteolysis of the anterior inferior vertebral endplate or the presence of a paravertebral soft-tissue abscess.[4] Involvement of the posterior elements is rare.[2,4] The infectious process may spread beneath the anterior longitudinal ligament to invade the neighboring vertebral body.[4,5] Involvement of two adjacent vertebral bodies in seen in 50% of cases of tuberculous osteomyelitis, whereas three or more adjacent vertebral bodies are involved in 25%.[4] In the remaining 25% of cases, the disease either is confined to one vertebral body or has spread to involve two or more noncontiguous vertebral bodies. The intervertebral disc space may narrow secondary to vertebral destruction with subsequent collapse of the disc into the vertebral body;[4] however, the disc space is maintained longer in tuberculosis than in pyogenic infection.[5] With continued anterior vertebral body destruction and maintenance of the posterior elements, kyphosis occurs.

CT has been useful in the evaluation of patients suspected of having tuberculosis of the spine. CT is more accurate than conventional radiography in demonstrating the extent of bone destruction and the presence of paravertebral or psoas abscess[1,2,5] (Figs. 22C–22E). Calcification within the abscess is readily detected with CT and reflects the chronicity of the tuberculous process.[3] The rim of the tuberculous abscess is thick and nodular. After the administration of intravenous contrast, enhancement of the rim may be demonstrated because of its hypervascularity.[5] When

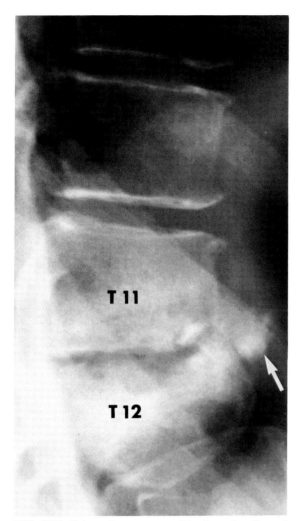

FIG. 22C. Tuberculous osteomyelitis. Lateral radiograph of the lower thoracic spine. There is narrowing of the T11-T12 intervertebral disc space. Destruction, sclerosis, and partial collapse of T12 are demonstrated. There is also partial destruction of the inferior margin of T11. Dense calcification or ossification is seen anteriorly *(arrow).*

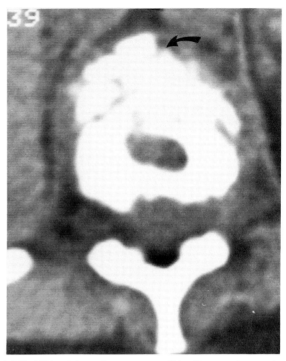

FIG. 22D. Tuberculous osteomyelitis. Same patient as in Fig. 22C. Axial CT at T11 demonstrates vertebral destruction. Ossification is present anteriorly *(arrow)* and is due to either bone fragmentation or calcified abscess. This 40-year-old woman had osseous tuberculosis of the lower leg at age 7, which has since been quiescent. She presented at this time with back pain. There was no neurologic deficit.

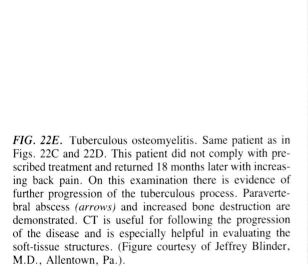

FIG. 22E. Tuberculous osteomyelitis. Same patient as in Figs. 22C and 22D. This patient did not comply with prescribed treatment and returned 18 months later with increasing back pain. On this examination there is evidence of further progression of the tuberculous process. Paravertebral abscess *(arrows)* and increased bone destruction are demonstrated. CT is useful for following the progression of the disease and is especially helpful in evaluating the soft-tissue structures. (Figure courtesy of Jeffrey Blinder, M.D., Allentown, Pa.).

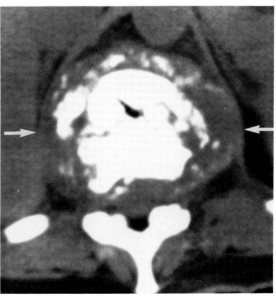

compared to a nontuberculous paraspinal abscess, a tuberculous paraspinal abscess is more likely to be multiloculated and calcified, to spread to superficial dorsal soft tissues, and to have a thick, irregular rim.[5] Neurologic symptoms may be caused by involvement of the paraspinal neural plexus or by epidural extension of the infectious process causing cord compression.[2] Epidural spread of infection can be evaluated by CT or CTM. CT is also useful in the guiding of biopsy procedures and as a follow-up examination for evaluation of therapeutic results.

References

1. Gropper GR, Acker JD, Robertson JH: Computed tomography in Pott's disease. *Neurosurgery* 1982;10:506–508.
2. LaBerge JM, Brant-Zawadzki M: Evaluation of Pott's disease with computed tomography. *Neuroradiology* 1984;26:429–434.
3. Maritz NGJ, deVilliers JFK, van Castricum OQS: Computed tomography in tuberculosis of the spine. *Comput Radiol* 1982;6:1–5.
4. Weaver P, Lifeso RM: The radiological diagnosis of tuberculosis of the adult spine. *Skeletal Radiol* 1984;12:178–186.
5. Whelan MA, Naidich DP, Post JD, et al: Computed tomography of spinal tuberculosis. *J Comput Assist Tomogr* 1983;7:25–30.

CASE 23

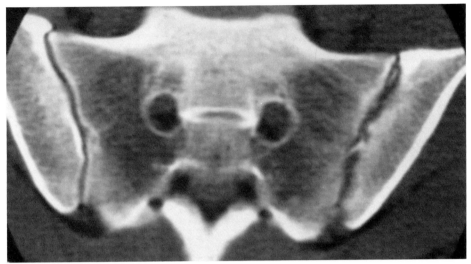

FIG. 23A. CT study of the sacroiliac joints. This 36-year-old patient is a drug addict who developed endocarditis and *Staphylococcus* septicemia. Vague back pain and continued fever despite antibiotic therapy led to a radionuclide bone scan, which revealed increased uptake in the left sacroiliac joint.

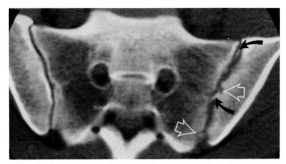

FIG. 23A-1. Pyogenic sacroiliitis. There is unilateral sacroiliitis with cortical irregularity and erosions on both sides of the left SI joint *(straight arrows)*. Small bony fragments are noted within the joint *(curved arrows)*.

Pyogenic Sacroiliitis

Pyogenic sacroiliitis may present as a subacute localized illness or an acute systemic disorder. The clinical presentation of buttock, hip, or abdominal pain in association with fever may be misleading and may suggest nerve root compression, a hip disorder, or an inflammatory process of the abdomen.[7] Septic arthritis of the sacroiliac (SI) joint may occur during pregnancy or immediately postpartum or postabortion; secondary to infection of the skin, bone, or urinary tract; or in association with drug abuse.[4] The early diagnosis of pyogenic sacroiliitis is important since it is associated with improved therapeutic prognosis. Infectious sacroiliitis (pyogenic or tuberculous) should always be considered in the differential diagnosis of unilateral sacroiliitis.

Radionuclide bone scanning is useful as a screening procedure for infectious diseases of the SI joint; however, it is a nonspecific examination. Occasionally, the bone scan may be falsely negative (e.g., normal SI joint activity may obscure appreciation of abnormal activity) or falsely positive (e.g., cellulitis with increased bone activity due to hyperemia rather than infection).[1,7] Conventional radiography is usually normal during the first 2 to 3 weeks following the onset of infection.[1,4] CT is more sensitive than conventional radiography in the detection of pyogenic sacroiliitis and can detect widening of the SI joint, cortical irregularity and destruction, and bone fragmentation within the joint[1,7,8] (Figs. 23A-1, 23B). The soft tissues are best visualized by CT. Soft-tissue extension of infection can be demonstrated as thickening of the iliac and gluteal muscles or as a paraarticular soft-tissue mass. Gas within a soft-tissue abscess is readily demonstrated with CT. If surgical

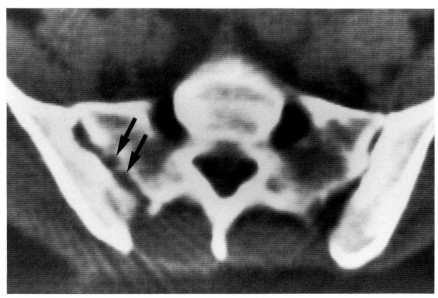

FIG. 23B. Pyogenic sacroiliitis. Cortical irregularity and erosions are present on the sacral side of the right SI joint *(arrows)*. Culture revealed *Staphylococcus aureus*. This 30-year-old woman presented with abdominal and back pain, which was initially believed to be due to an acute abdominal process.

intervention is required, CT is invaluable in precisely delineating the location and extent of the infectious process within and around the SI joint.[8] In addition, CT can be used as a guide for biopsy of infected paraarticular bone or for joint aspiration.

CT can also be used in the detection of noninfectious sacroiliitis associated with ankylosing spondylitis, Reiter's syndrome, psoriatic arthritis, ulcerative colitis, Crohn's disease, and gout (Figs. 23C, 23D). The radiographic evaluation of the SI joints plays a key role in the early diagnosis of the spondyloarthropathies. Radionuclide bone scanning is sensitive but nonspecific. Although CT is more sensitive than conventional radiography for the detection of sacroiliitis,[3,5] careful examination of good-quality radiographs is usually sufficient for proper diagnosis.[2,9] In one series, patients with low back pain that was clinically suspicious for sacroiliitis were exam-

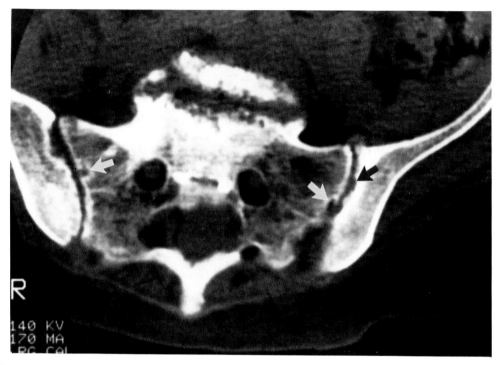

FIG. 23C. Sacroiliitis secondary to gout. Bilateral erosions are present on both the iliac and sacral sides of the SI joints *(arrows)*. This patient had gout involving multiple peripheral joints and presented at this time with a 2-year history of low back and left hip pain.

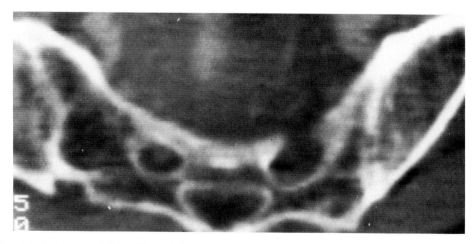

FIG. 23D. Ankylosing spondylitis. Bilateral fusion of the SI joints is present secondary to ankylosing spondylitis. At this advanced stage CT does not usually add significant information beyond that obtained from conventional radiographs.

ined by conventional radiography and CT.[9] The PA radiograph of the pelvis was interpreted unequivocally as either normal or abnormal in approximately 75% of cases. Two thirds of the remaining cases were diagnosed correctly when a complete four-view radiographic series was evaluated. Those patients with equivocal results despite a complete radiographic series (approximately 10%) were found to benefit from CT examination. Thus CT is useful in evaluating patients in whom a definitive diagnosis cannot be made by conventional radiography. CT may also be used when clinical suspicion is high despite normal radiographic findings.[9]

CT examination of the SI joints can be achieved with contiguous 4- or 5-mm-thick sections obtained from the midportion of the first sacral segment through the entire length of the synovial portion of the SI joint. Optimal scanning is achieved when the gantry is tilted as nearly parallel as possible to the sacrum. This requires a maximum angulation of 15° to 25°. The angulation is opposite the direction normally used for the L5-S1 intervertebral disc space (Fig. 23E). Using this gantry tilt, the synovial portion on the SI joint is differentiated from the ligamentous portion by its more ventral, caudal, and vertical position[6] (Figs. 23F–23H).

The SI joints of asymptomatic patients have been studied by CT, and the appearance of the SI joints has been found to vary with age.[10] In patients under the age of 30 years, the SI joints are uniform and

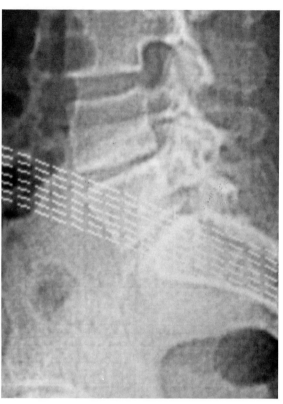

FIG. 23E. Lateral digital radiograph of the lumbar spine and sacrum with cursor lines positioned for ideal scanning of the SI joints. Maximum gantry angulation is used to scan as nearly parallel to the sacrum as possible. Note that this angulation is opposite that used to scan parallel to the L5-S1 disc.

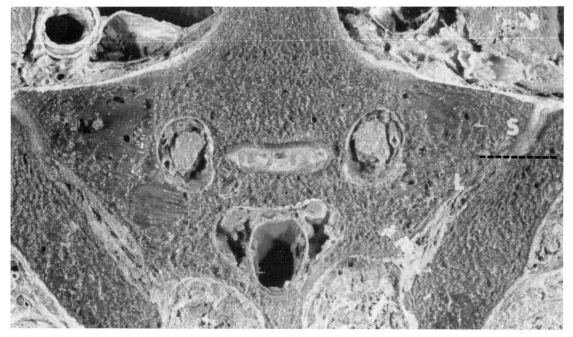

FIG. 23F. Normal SI joints. Anatomic specimen through the SI joints at the junction of the synovial and ligamentous portions. The synovial portion of the joint (*S*, above *dotted line*) is more ventral and vertical compared to the more dorsal and oblique ligamentous part of the joint (*L*, below *dotted line*).

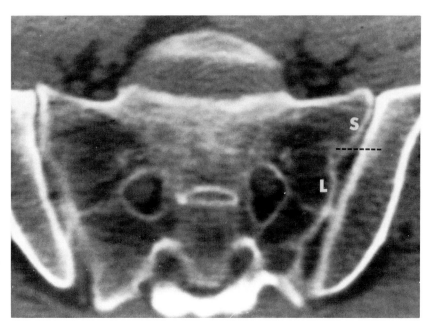

FIG. 23G. Normal SI joints. CT scan in a near-coronal plane. This scan is at the junction of the synovial (*S*, above *dotted line*) and ligamentous (*L*, below *dotted line*) portions of the SI joints, a similar plane to that shown in Fig. 23F. The ligamentous portion of the joint is more dorsal, oblique, and irregular than the synovial portion.

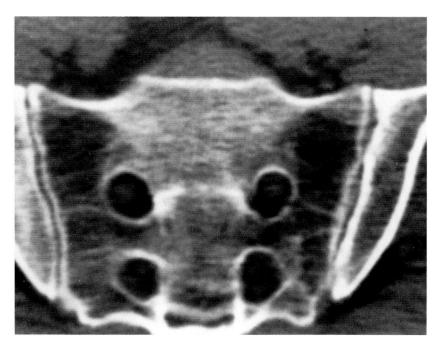

FIG. 23H. Normal SI joints. CT scan in a near-coronal plane. This scan is obtained caudad to Fig. 23G and is through the synovial portion of the SI joints, caudad to the ligamentous portion of the joints. Note the uniform symmetric appearance of the joints with no evidence of erosions, joint space narrowing, or other signs of sacroiliitis.

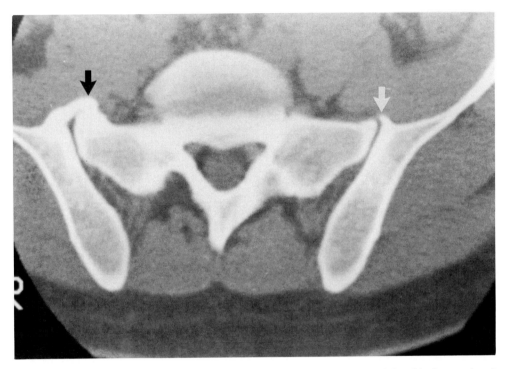

FIG. 231. Osteoarthritis of SI joints. Anterior para-articular ankylosis of the right SI joint *(black arrow)* and a small osteophyte of the left ilium *(white arrow)* represent degenerative changes not infrequently encountered in older age groups.

symmetric. It is not unusual to demonstrate iliac sclerosis that is focal and nonuniform. In the older age groups, asymptomatic patients may have focal joint space narrowing, focal iliac and sacral subchondral sclerosis that begins on the iliac side, para-articular ankylosis, and overall asymmetry of the joints[10] (Fig. 231). With these "normal aging" or degenerative findings in mind, diagnostic signs of sacroiliitis at any age would include *uniform* joint space narrowing (<2 mm), *intra-articular* ankylosis, and erosions.[3,5,10] Some CT findings that might be seen in asymptomatic older patients are suggestive of sacroiliitis when found in younger age groups. These findings include focal joint space narrowing, focal increased sacral subchondral sclerosis, and overall asymmetry of the joints.[10] Isolated iliac sclerosis is not considered a diagnostic finding of sacroiliitis.[3]

References

1. Bankoff MS, Sarno RC, Carter BL: CT scanning in septic sacroiliac arthritis or periarticular osteomyelitis. *Comput Radiol* 1984;8:165–170.
2. Borlaza GS, Seigel R, Kuhns LR, et al: Computed tomography in the evaluation of sacroiliac arthritis. *Radiology* 1981;139:437–440.
3. Carrera GF, Foley WD, Kozin F, et al: CT of sacroiliitis. *AJR* 1981;136:41–46.
4. Gordon G, Kabins SA: Pyogenic sacroiliitis. *Am J Med* 1980;69:50–56.
5. Kozin F, Carrera GF, Ryan LM, et al: Computed tomography in the diagnosis of sacroiliitis. *Arthritis Rheum* 1981;24:1479–1485.
6. Lawson TL, Foley WD, Carrera GF, et al: The sacroiliac joints: Anatomic, plain roentgenographic, and computed tomographic analysis. *J Comput Assist Tomogr* 1982;6:307–314.
7. Morgan GJ Jr, Schlegelmilch JG, Spiegel PK: Early diagnosis of septic arthritis of the sacroiliac joint by use of computed tomography. *J Rheumatol* 1981;8:979–982.
8. Rosenberg D, Baskies AM, Deckers PJ, et al: Pyogenic sacroiliitis: An absolute indication for computerized tomographic scanning. *Clin Orthop* 1984;184:128–132.
9. Ryan LM, Carrera GF, Lightfoot RW Jr, et al: The radiographic diagnosis of sacroiliitis: A comparison of different views with computed tomograms of the sacroiliac joint. *Arthritis Rheum* 1983;26:760–763.
10. Vogler JB III, Brown WH, Helms CA, et al: The normal sacroiliac joint: A CT study of asymptomatic patients. *Radiology* 1984;151:433–437.

CASE 24

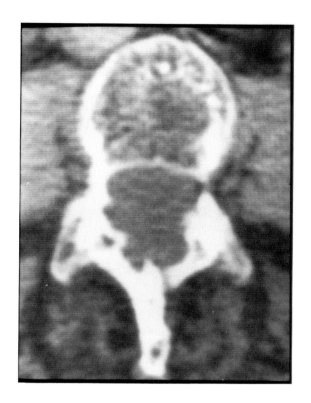

FIG. 24A. Axial CT scan at the superior aspect of the L3 vertebra. This 68-year-old woman had slowly progressive right leg weakness of several years' duration associated with atrophy of the right leg. She also had vague chronic back pain and stiffness.

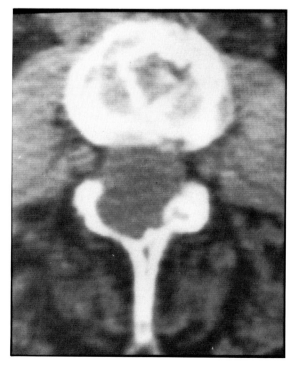

FIG. 24B. Axial CT at the inferior aspect of the L3 vertebra.

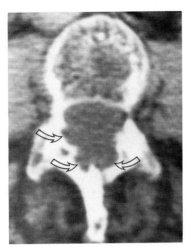 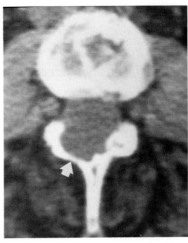

FIG. 24A-1. Ankylosing spondylitis. There are erosions of the inner margin of the laminae bilaterally *(arrows)* due to dorsal diverticula.

FIG. 24B-1. Similar scalloping of the laminae is noted, causing marked thinning of the lamina on the right *(arrow).* Note the moderate fatty replacement of the posterior paraspinal musculature.

Ankylosing Spondylitis

The CT examination demonstrates typical findings of ankylosing spondylitis with scalloping of the inner margin of the laminae and spinous processes at multiple lumbar levels[2,4,5] (Figs. 24A-1, 24B-1). Scalloping of the laminae is due to pressure erosion from dural ectasia and multiple dorsal diverticula.[7,8] The differential diagnosis of the CT finding of spinal canal erosion includes neurofibromatosis, Marfan's syndrome, tumors, and cysts.[2] Typically, patients with dural ectasia secondary to neurofibromatosis or Marfan's syndrome develop scalloping of the posterior vertebral margin, whereas patients with ankylosing spondylitis develop erosion of the posterior elements.[2] An extensive intraspinal tumor may have CT findings of scalloping, which are likely to involve both the laminae and the posterior vertebral bodies. In the presence of tumor the density measurements of the thecal sac may be increased. If there is uncertainty, myelography and CTM can be used to exclude the possibility of tumor. In patients with ankylosing spondylitis, contrast fills the posterior diverticula, extending into the eroded laminae.[7,8] Some patients with ankylosing spondylitis and CT findings similar to those in the present case have the cauda equina syndrome, with slowly progressive leg or buttock pain, sensory or motor impairment, and bowel or bladder dysfunction.[8]

An additional point of interest in this case is the fatty replacement of the posterior paraspinal musculature (Figs. 24A-1, 24B-1), which has been reported to occur in association with ankylosing spondylitis.[9] CT demonstration of very severe fatty replacement of paraspinal musculature has been re-

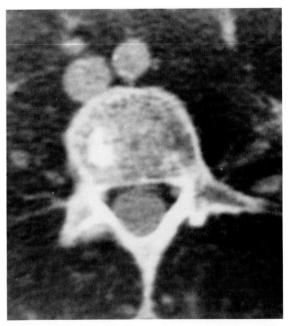

FIG. 24C. Fatty replacement of the paraspinal musculature. Axial CT scan demonstrates extensive fatty replacement of the paraspinal musculature in this patient with a 30-year history of paraparesis secondary to poliomyelitis.

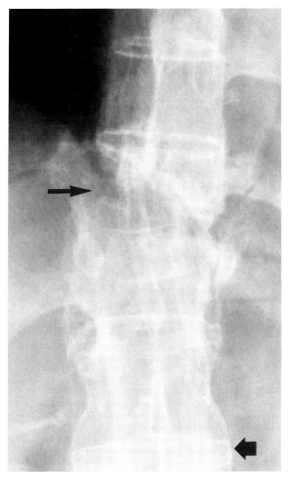

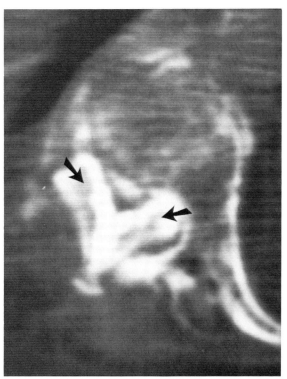

FIG. 24E. Transection fracture in a patient with ankylosing spondylitis. Same patient as in Fig. 24D. Axial CT scan dramatically demonstrates the fracture through the posterior elements. The laminae *(arrows)* and spinous process are displaced anteriorly with marked narrowing of the spinal canal and compression of the spinal cord.

FIG. 24D. Transection fracture of the thoracic spine in patient with ankylosing spondylitis. This patient sustained a severe fracture of T11 *(long arrow)* after a fall on stairs. The asymmetry of the spinous processes indicates a rotatory component of the fracture. Note that syndesmophytes *(short arrow)* of ankylosing spondylitis are present.

ported in patients with neuromuscular disorders such as poliomyelitis and muscular dystrophy[3] (Fig. 24C). Fatty replacement of the sacrospinal muscle groups may also be seen in varying degrees of severity in patients who have had previous lumbar surgery[6] and in otherwise normal individuals, especially elderly females.[3]

Patients with ankylosing spondylitis are more susceptible to fracture-dislocation secondary to minor trauma than are normal individuals. The spine in patients with ankylosing spondylitis is rigid and tends to fracture as does a long bone, with complete through-and-through fractures rather than the more common vertebral compression fractures. The fractures tend to cross the intervertebral disc spaces to involve adjacent vertebrae, and fractures of the posterior elements are frequent. These fractures are associated with a high morbidity and mortality. In one study, 12% of patients with ankylosing spondylitis had a spinal fracture, and 8% of this group had paralytic spinal cord injury.[10] Spinal injuries in patients with ankylosing spondylitis are frequently unstable and have associated dislocation, neural arch displacement, or complete transection.[1,10] (Figs. 24D, 24E). CT may demonstrate neural arch fractures and ventral displacement of fracture fragments not identified by conventional radiography.[10] The degree of compression of the spinal cord or nerve roots can readily be evaluated by CT. Spinal cord contusion without evidence of fracture has also been described.[10]

Other radiographic features of ankylosing spondylitis include bilateral symmetric sacroiliitis and spondylitis with thin syndesmophytes; squaring of vertebrae; fusion of apophyseal joints; calcification or ossification of interspinous and supraspinous ligaments; intervertebral disc calcification; and, rarely, atlantoaxial subluxation.

References

1. Grisolia A, Bell RL, Peltier LF: Fractures and dislocations of the spine complicating ankylosing spondylitis. *J Bone Joint Surg Am* 1967;49-A:339–344, 386.
2. Grosman H, Gray R, St Louis EL: CT of long-standing ankylosing spondylitis with cauda equina syndrome. *AJNR* 1983;4:1077–1080.
3. Hadar H, Gadoth N, Heifetz M: Fatty replacement of lower paraspinal muscles: Normal and neuromuscular disorders. *AJNR* 1983;4:1087–1090, *AJR* 1983;141:895–898.
4. Helms CA, Vogler JB: Computed tomography of spinal stenoses and arthroses. *Clin Rheum Dis* 1983;9:417–441.
5. Kramer LD, Krouth GJ: Computerized tomography: An adjunct to early diagnosis in the cauda equina syndrome of ankylosing spondylitis. *Arch Neurol* 1978;35:116–118.
6. Laasonen EM: Atrophy of sacrospinal muscle groups in patients with chronic, diffusely radiating lumbar back pain. *Neuroradiology* 1984;26:9–13.
7. Rosenkranz W: Ankylosing spondylitis: Cauda equina syndrome with multiple spinal arachnoid cysts. *J Neurosurg* 1971;34:241–243.
8. Russell ML, Gordon DA, Ogryzlo MA, et al: The cauda equina syndrome of ankylosing spondylitis. *Ann Intern Med* 1973;78:551–554.
9. Sage MR, Gordon TP: Muscle atrophy in ankylosing spondylitis: CT demonstration. *Radiology* 1983;149:780.
10. Weinstein PR, Karpman RR, Gall EP, et al: Spinal cord injury, spinal fracture and spinal stenosis in ankylosing spondylitis. *J Neurosurg* 1982;57:609–616.

CASE 25

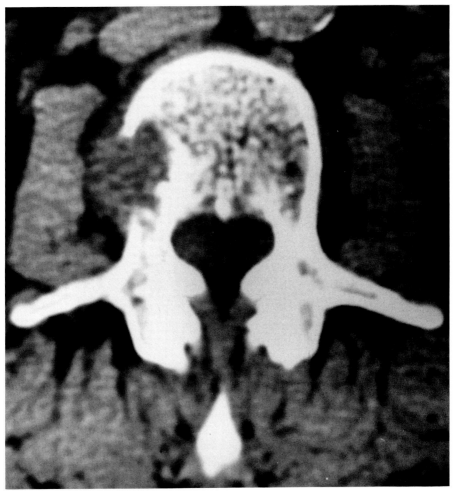

FIG. 25A. CT scan at L3 in a 69-year-old woman with right thigh and hip pain.

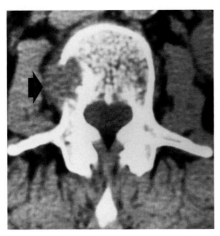

FIG. 25A-1. Osteolytic metastasis to the vertebral body from carcinoma of the breast. A destructive lesion of the right side of the vertebral body has broken through the cortex and is extending into the paravertebral soft tissues *(arrow)*.

Vertebral Metastasis

There is a solitary osteolytic lesion destroying the right side of the vertebral body (Fig. 25A-1). The lesion is poorly defined and extends into the paravertebral soft tissues. This represents metastasis from carcinoma of the breast.

Metastasis is the most common tumor of the spine. It may occur at any age but is progressively more frequent after the age of 40 as the incidence of primary tumors rises.[5] Hematogenously spread metastasis to the spine is most often from carcinoma of the breast, prostate, or lung in adults and from neuroblastoma in the infant and child. Initially, metastasis usually attacks the vertebral body, although pedicles may also be involved early, either by hematogenous spread or by direct extension of tumor from the vertebral body. The axial skeleton is a frequent site for metastatic deposits because of its rich red marrow distribution. The thoracic and lumbosacral regions of the spine are most often affected. Although most metastatic disease to the spine spreads hematogenously, direct extension from a paravertebral tumor can occur and cause destruction or scalloping of the vertebral body margin (Fig. 25B).

In the workup of patients suspected of having metastatic disease to the spine, radionuclide scanning is used as a screening procedure because of its high sensitivity. However, the radionuclide scan lacks specificity and frequently is positive in response to benign disorders.[6] The conventional radiograph follows the radionuclide scan in an effort to distinguish metastatic lesions from benign disorders. However, the radiograph lacks sensitivity, and approximately 50% of cancellous bone must have been destroyed before spinal metastasis can be detected.[2] CT can detect osseous metastasis in the spine more readily than conventional radiography because of its superior resolution.[7,8] CT makes an important contribution in the evaluation of patients with primary disease at high risk for bone metastasis (i.e., carcinoma of the breast, lung, prostate) who have a positive radionuclide bone scan and normal conventional radiographs.[3,4,7,8] In one study of 20 such patients with positive bone scan and normal radiographs, CT detected spinal metastatic disease in 80% of the cases and demonstrated degenerative changes as the cause of the positive bone scan in the remaining cases.[8]

Metastasis to the spine may appear on CT examination as single or multiple osteolytic lesions. The margin of the metastatic lesion usually lacks sclerosis. The tumor may be focal or may be widespread, involving the entire vertebral body. Metastasis may also appear osteoblastic, a finding that is frequent with metastasis from carcinoma of the prostate gland and lymphoma and may also be seen in patients with breast carcinoma (Fig. 25C). The sclerosis is not tumor but rather reactive (normal) bone formation due to the presence of underlying tumor.[5] An osteolytic metastatic lesion may become osteosclerotic following radiation or chemotherapy as tumor growth is slowed or arrested. Mixed osteolytic and osteoblastic lesions may also be noted on CT. Paravertebral and epidural extension of vertebral metastasis can be demonstrated.

The use of CT for the early detection of spinal metastasis is important clinically since it can lead to appropriate therapy, thereby preventing vertebral fracture and/or epidural extension of tumor with subsequent spinal cord or cauda equina compression.[8] In addition, CT can be used to guide closed-needle biopsy of vertebral and paravertebral tumor, thus diminishing the risk of complications associated with closed-needle biopsy performed with conventional radiographic guidance[1] (Fig. 25D). CT is also helpful in evaluating the extent of tumor before and after radiotherapy.

The differential diagnosis of metastasis to the spine includes multiple myeloma and chordoma in the adult and histiocytosis and leukemia in the child. A Schmorl's node (intraosseous herniation of the nucleus pulposus) can be differentiated from a metastatic lesion of the vertebral body by its location at the vertebral body endplate and by the presence of a sclerotic rim (Fig. 25E).

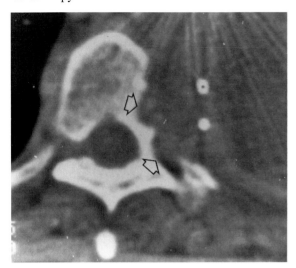

FIG. 25B. Metastasis from direct extension of tumor. This 65-year-old patient had a left pneumonectomy for carcinoma of the lung and presented at this time with increasing back pain. Recurrent tumor has spread to the paravertebral soft tissues and has invaded the bone, causing scalloping of the margins of the T6 vertebral body and posterior elements *(arrows)*. Note the postpneumonectomy state.

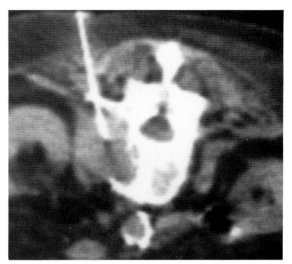

FIG. 25D. Closed-needle biopsy performed under CT guidance. The patient is in the prone position. There is destruction of the left side of the L3 vertebral body with paravertebral extension of tumor. A needle biopsy demonstrated the presence of metastatic adenocarcinoma. The use of CT for closed-needle biopsy guidance represents another important application of CT.

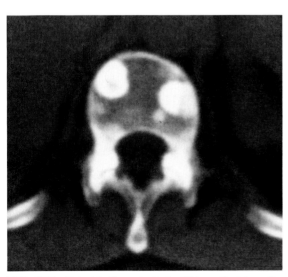

FIG. 25C. Osteoblastic metastasis. This patient has discrete osteoblastic lesions to the vertebral body, pedicles, and articular processes from carcinoma of the prostate gland.

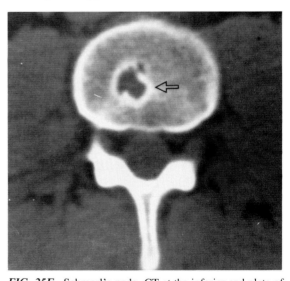

FIG. 25E. Schmorl's node. CT at the inferior end plate of L3 demonstrates a low-density Schmorl's node (intraosseous herniation of the nucleus pulposus) with a surrounding sclerotic rim *(arrow)*. The location at the endplate and the presence of a sclerotic rim differentiate this from metastatic disease.

References

1. Adapon BD, Legada BD Jr, Lim EVA, et al: CT-guided closed biopsy of the spine. *J Comput Assist Tomogr* 1981;5:73–78.
2. Edelstyn GA, Gillespie PJ, Grebbell FS: The radiological demonstration of osseous metastases: Experimental observations. *Clin Radiol* 1967;18:158–162.
3. Harbin WP: Metastatic disease and the nonspecific bone scan: Value of spinal computed tomography. *Radiology* 1982;145:105–107.
4. Helms CA, Cann CE, Brunelle FO, et al: Detection of bone-marrow metastases using quantitative computed tomography. *Radiology* 1981;140:745–750.
5. Kricun ME: Radiographic evaluation of solitary bone lesions. *Orthop Clin North Am* 1983;14:39–63.
6. Mall JC, Bekerman C, Hoffer PB, et al: A unified approach to the detection of skeletal metastases. *Radiology* 1976;118:323–328.
7. Muindi J, Coombes RC, Golding S, et al: The role of computed tomography in the detection of bone metastases in breast cancer patients. *Br J Radiol* 1983;56:233–236.
8. Redmond J III, Spring DB, Munderloh SH, et al: Spinal computed tomography scanning in the evaluation of metastatic disease. *Cancer* 1984;54:253–258.

CASE 26

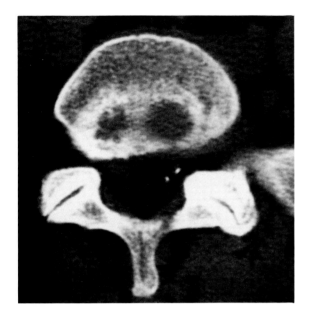

FIG. 26A. Axial CT at the inferior aspect of the L5 vertebral body, 5 mm cephalad to the L5–S1 disc.

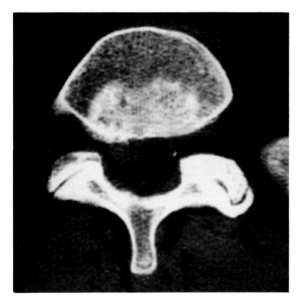

FIG. 26B. Axial CT at L5, 5 mm cephalad to Fig. 26A.

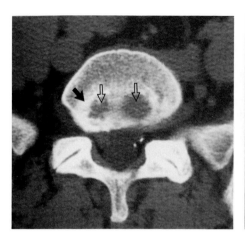 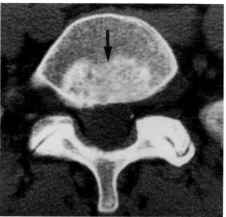

FIG. 26A-1. "Cupid's bow" contour. There are paired parasagittal round areas of low density (*open arrows*) with surrounding sclerosis (*closed arrow*). The CT appearance is that of "owl's eyes" and is due to a normal variant termed "Cupid's bow." This structure is typically located in the posterior inferior aspect of a lower lumbar vertebral body.

FIG. 26B-1. Intense sclerosis (*arrow*) is present cephalad to the parasagittal areas of low density and is another typical finding of the Cupid's bow contour. This should not be confused with osteoblastic metastatic disease.

"Cupid's Bow"

The scan obtained at the inferior subchondral aspect of the vertebra reveals two round areas of lower density surrounded by sclerosis (Fig. 26A-1). The adjacent higher scan demonstrates intense sclerosis (Fig. 26B-1). These are the typical CT findings of a normal variant known as the "Cupid's bow" contour. This appearance should not be confused with osteolytic or osteoblastic metastasis. The name "Cupid's

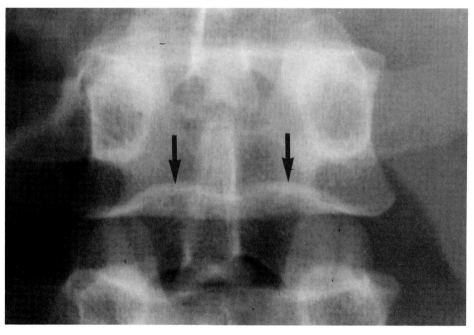

FIG. 26C. Cupid's bow contour. This AP radiograph of the lumbar spine demonstrates the Cupid's bow contour (*arrows*) with parasagittal indentations of the inferior endplate.

bow" is given to the appearance of paired parasagittal concavities of the inferior end plates of lumbar vertebrae as visualized on AP radiographs (Fig. 26C).

The process that leads to this anatomic deformity is not known but is thought to be related to the turgor of the nucleus pulposus as it expands against the cartilaginous endplate.[4] The concept that Cupid's bow is formed as a result of a variation in notochord development is not as plausible since the notochord is a midline structure, and the Cupid's bow contour impressions are parasagittal in location.[1,2]

The inferior aspects of the L3, L4, and L5 vertebrae are most commonly involved.[1] Rarely, a similar radiographic appearance occurs in the superior aspect of the vertebral body.[3] The CT scan demonstrates paired parasagittal areas of disk density located posteriorly within 5 to 7 mm of the inferior cartilaginous endplate.[2] Apparent sclerosis surrounding the radiolucencies is due to depressed subchondral bone and creates an "owl's eyes" appearance on CT.[3] Additional sclerosis is seen at the apex of the Cupid's bow.

The typical CT features and location of Cupid's bow should prevent confusion with metastatic disease. Osteolytic metastasis does not usually appear as paired parasagittal radiolucencies with surrounding sclerotic margins in the subchondral location. Intraosseous herniation of disc tissue (Schmorl's node) is located in subchondral bone and has surrounding sclerosis; however, unlike Cupid's bow it does not have a paired, parasagittally symmetric appearance.

References

1. Dietz GW, Christensen EE: Normal "Cupid's bow" contour of the lower lumbar vertebrae. *Radiology* 1976;121:577–579.
2. Firooznia H, Tyler I, Golimbu C, et al: Computerized tomography of the Cupid's bow contour of the lumbar spine. *Comput Radiol* 1983;7:347–350.
3. Ramirez H Jr, Navarro JE, Bennett WF: "Cupid's bow" contour of the lumbar vertebral endplates detected by computed tomography. *J Comput Assist Tomogr* 1984;8:121–124.
4. Resnick D, Niwayama G: Intravertebral disk herniations: Cartilaginous (Schmorl's) nodes. *Radiology* 1978;126:57–65.

CASE 27

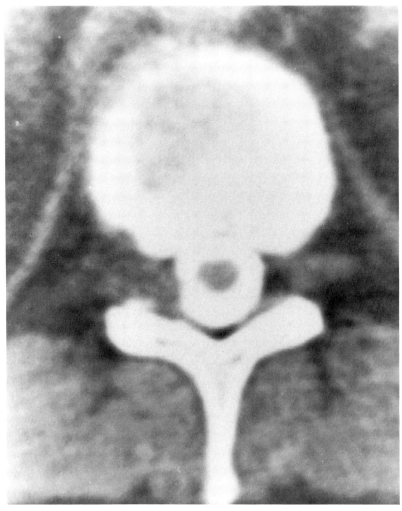

FIG. 27A. CTM at T11 in a 40-year-old male with a 3-year history of mediastinal teratocarcinoma and a 2-week history of increasing back pain and leg weakness.

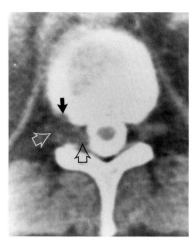

FIG. 27A-1. Epidural metastasis from cystic teratocarcinoma. There is a mass in the epidural space and neural foramen on the right (*open arrows*). Note the obliteration of the epidural fat compared to the normal left side. The contrast-filled subarachnoid space is slightly flattened by the adjacent tumor. The irregularity of the posterolateral margin of the vertebral body is due to direct extension of the tumor (*closed arrow*).

Epidural Metastasis

The contrast-filled thecal sac is asymmetric, the right lateral margin being flattened by an epidural mass (Fig. 27A-1). There is slight irregularity of the adjacent vertebral body due to direct extension of the tumor. The epidural mass was metastatic teratocarcinoma. Isolated metastasis to the epidural space is found in 5% of patients with spinal metastasis presenting with signs of spinal cord or cauda equina compression.[2] Much more frequently, epidural metastasis develops from direct extension of osseous vertebral metastasis, usually secondary to carcinoma of the breast, lung, or prostate gland.[2,5] Patients with known or suspected primary carcinoma who have back pain and a myelopathy, a radiculopathy, or radiographic evidence of metastasis at the appropriate clinical level need further evaluation for possible epidural metastasis.[5]

Myelography[5] and CT[9] have both been utilized in the evaluation of patients with primary malignancy and clinical suspicion of spinal cord or nerve root compression. Myelography permits the demonstration of a partial or complete block and allows examination of the entire spinal canal, which may reveal other subclinical intraspinal metastases. Although CT cannot practically be used to evaluate the entire spinal canal, it is nevertheless a highly accurate method of detecting epidural tumor causing spinal cord or nerve root compression.[9] In addition, the CT examination can better delineate osseous metastatic disease as well as paravertebral and retroperitoneal extension of tumor. Conventional CT, without intrathecal contrast, may demonstrate osseous metastatic disease with extension of tumor into the epidural space. The cortical margins of bone are destroyed, and soft-tissue extension of tumor causes obliteration of epidural fat (Fig. 27B). However, it may be difficult to detect small epidural metastatic foci, particularly in the cervical and thoracic spine where there is normally a deficiency of epidural fat. CTM has an advantage over unenhanced CT in that epidural and the less frequent intradural metastasis can be detected more readily, particularly if no associated bony metastasis is present or if the epidural tumor is small[1,5,7,9] (Fig. 27A-1). An epidural tumor appears on CTM examination as a mass of soft-tissue density causing compression of the subarachnoid space and displacement of the spinal cord or cauda equina (Figs. 27C–27G). If a myelogram is performed with water-soluble contrast and a block is demonstrated, CTM can often determine the extent of tumor beyond the block (Figs. 27H, 27I), thus providing information that is vitally important if radiation therapy is needed.[3,4,7,8]

Occasionally tumor from a paraspinal mass extends into the spinal canal (Fig. 27J). The effect of the invading tumor on the subarachnoid space and spinal cord can be determined by CTM. Intravenous injection of iodinated contrast usually causes some enhancement of metastatic tumor and may occasionally demonstrate tumor infiltration of the epidural space that was not appreciated on the precontrast scan.[6]

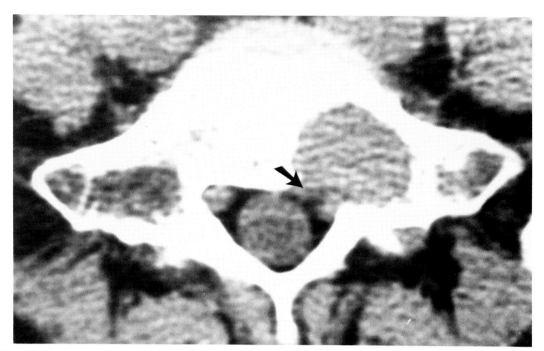

FIG. 27B. Epidural extension of osseous metastasis. This patient has metastasis to the sacrum from carcinoma of the breast. The axial CT study performed without intrathecal contrast demonstrates a destructive lesion of the sacrum with extension through the posterior cortex and into the epidural space (*arrow*). Note the partial obliteration of the anterior epidural fat on the left and displacement of the S1 nerve root.

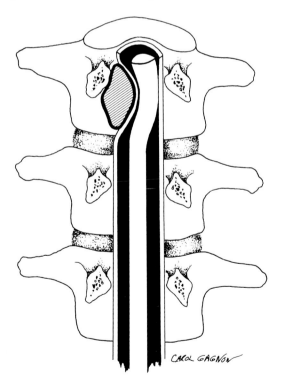

FIG. 27C. Relationship of an extradural tumor (*shaded area*) to the subarachnoid space (*black area*) and spinal cord in the coronal plane. This type of tumor causes displacement and compression of the subarachnoid space and the spinal cord.

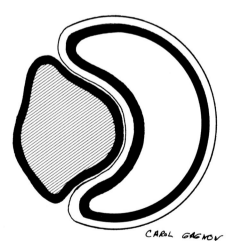

FIG. 27D. Relationship of an extradural tumor (*shaded area*) to the subarachnoid space and spinal cord in the axial plane. The effects of an extradural tumor on the subarachnoid space and spinal cord are similar to those seen with axial CTM.

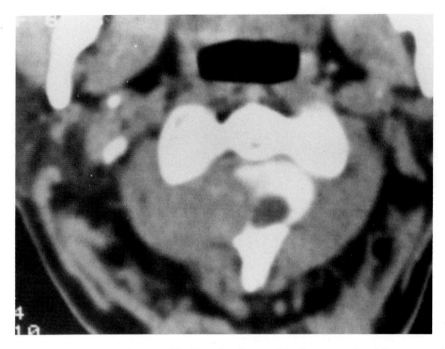

FIG. 27E. Epidural melanoma. Axial CTM at C2. There is a large epidural mass on the right compressing the contrast-filled subarachnoid space and displacing the spinal cord to the left. Compare with Fig. 27D. This 66-year-old female had a 3-month history of posterior neck pain and numbness, paresthesias, and weakness of the extremities. No other site of melanoma was found, and this is thought to represent a primary epidural melanoma.

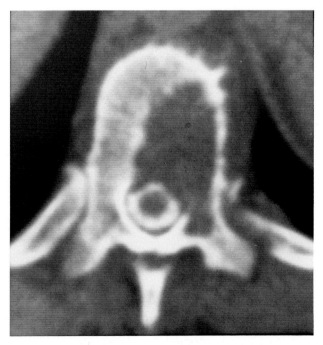

FIG. 27F. Osseous metastasis with epidural extension. Axial CTM examination at T11. This patient has metastatic hypernephroma. There is osseous destruction of the vertebral body, left pedicle, and lamina. Tumor has spread to the epidural space, causing thinning of the anterior contrast-filled subarachnoid space.

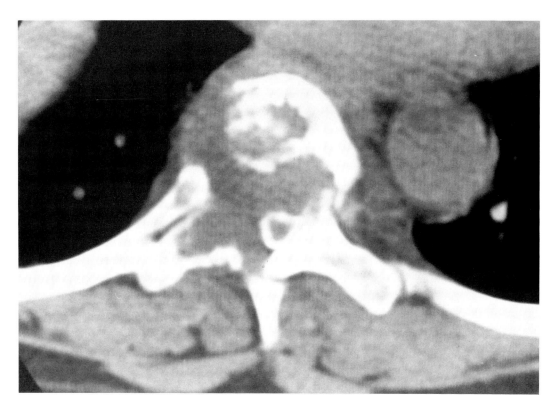

FIG. 27G. Osseous metastasis with epidural extension and pathologic fractures. Axial CTM demonstrates destruction of the vertebral body, pedicles, transverse process, and lamina. Pathologic fractures of the pedicles and lamina are seen.

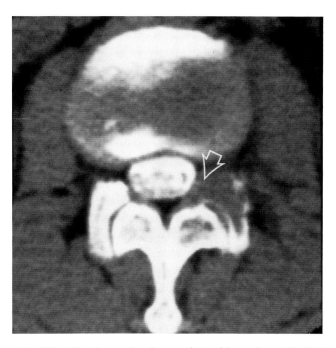

FIG. 27H. Osseous metastasis with epidural extension in a patient with myelographic block at L2. Intrathecal contrast was introduced at a lower lumbar level. Axial CTM below the level of the block demonstrates metastasis to the superior articular process on the left with extension of tumor to the epidural space (*arrow*) abutting the contrast-filled sac.

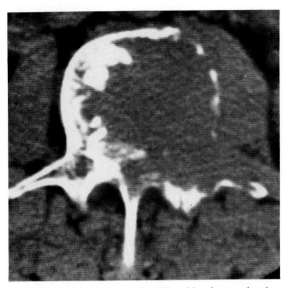 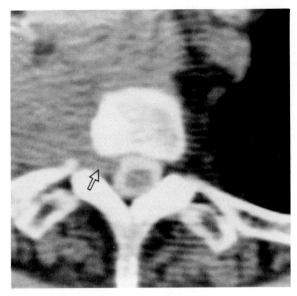

FIG. 27I. Osseous metastasis with epidural extension in a patient with myelographic block. Same patient as in Fig. 27H. CTM at L2, the level of myelographic block. No intrathecal contrast is seen; however, the extensive tumor involvement of the osseous and soft-tissue structures can be appreciated.

FIG. 27J. Direct extension of bronchogenic carcinoma into the neural foramen and epidural space (*arrow*). Axial CTM at T2. There is subtle flattening of the contrast-filled subarachnoid space on the right secondary to extension of apical lung carcinoma.

References

1. Anand AK, Krol G, Deck MDF: Lumbosacral epidural metastases: CT evaluation and comparison with myelography. *Comput Radiol* 1983;7:351–354.
2. Constans JP, DeDivitiis E, Donzelli R, et al: Spinal metastases with neurological manifestations: Review of 600 cases. *J Neurosurg* 1983;59:111–118.
3. Fink IJ, Garra BS, Zabell A, et al: Computed tomography with metrizamide myelography to define the extent of spinal canal block due to tumor. *J Comput Assist Tomogr* 1984;8:1072–1075.
4. Resjö IM, Harwood-Nash DC, Fitz CR, et al: CT metrizamide myelography for intraspinal and paraspinal neoplasms in infants and children. *AJR* 1979;132:367–372.
5. Rodichok LD, Harper GR, Ruckdeschel JC, et al: Early diagnosis of spinal epidural metastases. *Am J Med* 1981;70:1181–1188.
6. Schubiger O, Valavanis A, Hollmann J: Computed tomography of the intervertebral foramen. *Neuroradiology* 1984;26:439–444.
7. Tadmor R, Cacayorin ED, Kieffer SA: Advantages of supplementary CT in myelography of intraspinal masses. *AJNR* 1983;4:618–621.
8. Tan WS, Wilbur AC, Spigos DG: Postmyelographic CT evaluation of multiple blocks due to metastases: Case report. *J Comput Assist Tomogr* 1985;9:979–981.
9. Wang AM, Lewis ML, Rumbaugh CL, et al: Spinal cord or nerve root compression in patients with malignant disease: CT evaluation. *J Comput Assist Tomogr* 1984;8:420–428.

CASE 28

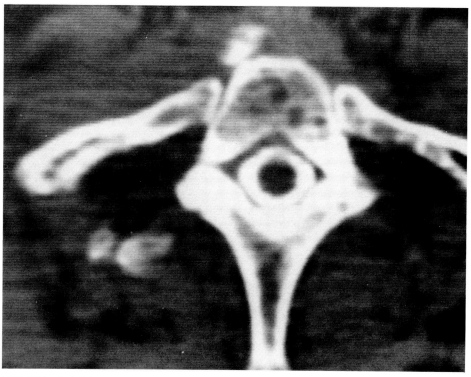

FIG. 28A. Axial CTM at T1. This 67-year-old male had right arm pain for 5 months with bilateral leg weakness. What diagnosis would you consider?

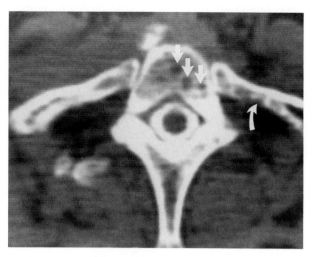

FIG. 28A-1. Multiple myeloma. There are multiple small lytic lesions of the T1 vertebral body (*straight arrows*). Additional destruction is seen in the spinous process and the left rib (*curved arrow*). There is no epidural extension at this level.

Myeloma

There are multiple small round lytic lesions within the vertebral body, spinous process, and rib (Fig. 28A-1). These are due to multiple myeloma, a malignant proliferation of plasma cells. Multiple myeloma is, after metastasis, the second most frequent malignancy of the spine. The CT appearance of multiple myeloma is variable. Osteolytic lesions occur with involvement of one or more vertebrae. The osteolytic pattern may consist of multiple small foci measuring 1 to 5 mm in diameter (Fig. 28A-1); however, lesions greater than 10 mm are frequent.[2,7] Destruction is sometimes so extensive that vertebral collapse occurs, and tumor or hemorrhage extends into the paravertebral and epidural spaces causing spinal cord or cauda equina compression (Fig. 28B). Myeloma may produce an expansile lesion that encroaches on the spinal canal and neural elements (Figs. 28C, 28D). Rarely, myeloma presents with osteosclerotic bone lesions, which may occur in untreated patients or may develop following radiation therapy or chemo-

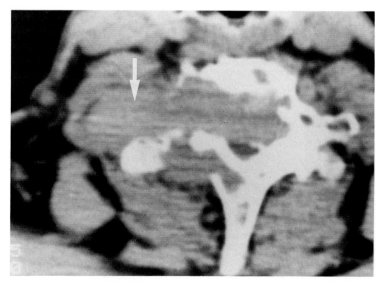

FIG. 28B. Multiple myeloma. Same patient as in Fig. 28A-1. Axial CTM at C7. There is extensive destruction of the C7 vertebral body, right pedicle, lamina, and transverse process. A soft-tissue mass extends into the epidural space, causing complete CTM block. A large right paravertebral mass is also present (*arrow*). The extensive osseous, epidural, and paravertebral involvement is the cause of the patient's symptoms of arm pain and leg weakness.

therapy. Osteosclerotic myeloma may also be associated with a syndrome of polyneuropathy, organomegaly, endocrinopathy, M protein, and skin changes: the so-called POEMS syndrome.[5] Myeloma frequently attacks the spine since the vertebral bodies are rich in red marrow. The pedicles, which possess little red marrow, are thus infrequently involved early in the course of the disease. However, the pedicles may be destroyed with more advanced disease as fat marrow in the pedicles reconverts to red marrow, subjecting the pedicles to the development of myeloma.[3]

Early in its course, myeloma is not readily detected with conventional radiography since it takes approximately 50% destruction of cancellous bone before a lesion is radiographically visible.[1] Nevertheless, conventional radiography is still more sensitive in detecting myeloma than is the radionuclide scan.[8] Radionuclide scanning, which is extremely sensitive to the presence of metastasis, is normal in 27% of patients with radiographically proven myeloma.[8] CT is more sensitive than conventional radiography in discovering vertebral myeloma and may detect lesions when conventional radiographs and radionuclide scans are unrewarding.[2,6,7] The CT findings are usually more extensive than expected. CT should not be used as a screening procedure for investigating the presence of myeloma in the entire spine; however, CT may be utilized for the detection of myeloma in a select group of patients. CT can adequately demonstrate myeloma in patients who have questionable conventional radiographic findings. It can also be utilized in patients who are clinically suspected of having myeloma (back pain, anemia, abnormal plasma protein electrophoresis, and proteinuria) yet have normal conventional radiographs and radionuclide scans. In these patients, CT scanning can be performed to include the vertebrae in the symptomatic regions. In patients with multiple myeloma, bone pain, and normal radiographs, CT evidence of bony involvement at the symptomatic site often indicates a need to initiate therapy.[6] Neurologic symptoms are related to compression of neural tissues by either vertebral collapse or extension of tumor into the epidural space.[4] CTM is indicated in those patients who have clinical evidence of spinal cord or cauda equina compression.

The CT differential diagnosis includes metastasis and severe osteoporosis. Metastasis may appear similar to all radiographic forms of myeloma. Severe osteoporosis, which may have mottled radiolucency in the vertebral body, does not appear destructive on CT as does myeloma. Also, the presence of a normal trabecular pattern in other vertebrae rules against diffuse osteoporosis.[2]

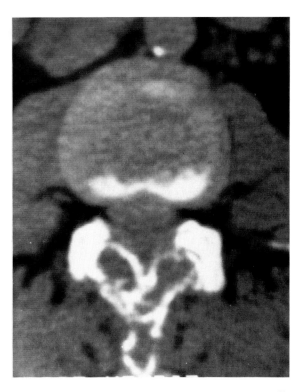

FIG. 28C. Myeloma. Axial CT at L2. There is expansile lytic destruction of the laminae and spinous process. This 55-year-old male presented with weight loss and a 2-month history of low back pain radiating into both legs. No other skeletal involvement was found. This proved to be the initial lesion of myeloma in a patient in whom multiple sites of involvement subsequently developed.

FIG. 28D. Myeloma. Same patient as in Fig. 28C. Sagittal reconstruction of axial CTM images demonstrating a partial block of contrast. Intrathecal contrast (*arrows*) is seen both above and below L2 but is not present at the involved L2 level. Destruction of the spinous process is again noted. *A*, anterior; *P*, posterior.

References

1. Edelstyn GA, Gillespie PJ, Grebbell FS: The radiological demonstration of osseous metastases: Experimental observations. *Clin Radiol* 1967;18:158–162.
2. Helms CA, Genant HK: Computed tomography in the early detection of skeletal involvement with multiple myeloma. *JAMA* 1982;248:2886–2887.
3. Kricun ME: Red-yellow marrow conversion: Its effect on the location of some solitary bone lesions. *Skeletal Radiol* 1985;14:10–19.
4. Loftus CM, Michelsen CB, Rapoport F, et al: Management of plasmacytomas of the spine. *Neurosurgery* 1983;13:30–36.
5. Resnick D, Greenway GD, Bardwick PA, et al: Plasma-cell dyscrasia with polyneuropathy, organomegaly, endocrinopathy, M-protein, and skin changes: The POEMS syndrome. *Radiology* 1981;140:17–22.
6. Schreiman JS, McLeod RA, Kyle RA, et al: Multiple myeloma: Evaluation by CT. *Radiology* 1985;154:483–486.
7. Solomon A, Rahamani R, Seligsohn U, et al: Multiple myeloma: Early vertebral involvement assessed by computerized tomography. *Skeletal Radiol* 1984;11:258–261.
8. Woolfenden JM, Pitt MJ, Durie BGM, et al: Comparison of bone scintigraphy and radiography in multiple myeloma. *Radiology* 1980;134;723–728.

CASE 29

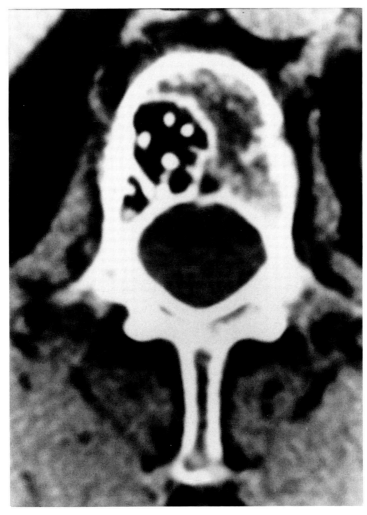

FIG. 29A. Axial CT scan at the level of L1.

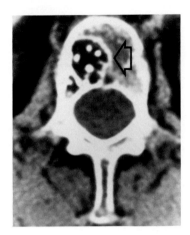

FIG. 29A-1. An osteolytic lesion is present within the vertebral body (*arrow*). The thick sclerotic margin indicates slow growth. The matrix of the lesion measured −74 HU. Several rounded densities are noted within the lesion, representing thickened trabeculae. This is the characteristic appearance of hemangioma.

Hemangioma

There is a sharply marginated osteolytic lesion of the vertebral body with thickened trabeculae that appear as multiple round densities within the matrix (Fig. 29A-1). This is a characteristic CT appearance of the so-called "hemangioma," the most frequent benign "tumor" of the vertebral body. Actually vertebral hemangiomas are not true tumors since they consist of dilated engorged blood vessels (venules and capillaries) reflecting venous stasis.[3] These dilated venous channels cause resorption of numerous trabeculae, so that the remaining vertical trabeculae thicken in response to mechanical stress. Hemangioma occurs in about 10% of the population, is single in two thirds of the cases, and is most frequent in the thoracic and lumbar spine.[5] It usually develops in the vertebral body and may extend into the posterior elements. Occasionally, hemangioma is isolated to the vertebral arch. Most patients are asymptomatic, and the lesion is an incidental finding of conventional radiography or CT. Rarely, patients become symptomatic because of compression of the spinal cord, cauda equina, or nerve root.[1,2,7] This usually occurs when hemangioma causes expansion of posterior elements leading to central spinal stenosis or lateral recess stenosis. Neural compression may also be caused by extension of hemangioma into the epidural space, compression fracture, or hemorrhage.[1,5] It is possible that some of the reported symptomatic lesions are true vascular tumors,[5] such as benign angioma, malignant hemangioendothelioma (angiosarcoma), or the rare hemangiopericytoma.[6]

No further imaging studies are necessary when hemangioma is discovered on the conventional radiograph of an asymptomatic patient. The classic honeycomb trabecular pattern is easily recognized with conventional radiography. Hemangioma can be accurately diagnosed by CT when conventional radiography is equivocal. CT examination demonstrates a well-defined osteolytic lesion. Multiple round densities representing thickened trabeculae are detected within the matrix and form the characteristic polkadot pattern. Hemangiomas are known to contain fat tissue within the matrix,[4] and we have observed negative attenuation values on CT scan indicating fat tissue. This may serve as an additional clue to the diagnosis of hemangioma.

In symptomatic patients, CT (and CTM when necessary) can detect the paravertebral or epidural extent of hemangioma.[2,7] Should intravenous contrast be administered, the enhancement of hemangioma attests to its marked vascularity.[7]

References

1. Healy M, Herz DA, Pearl L: Spinal hemangiomas. *Neurosurgery* 1983;13:689–691.
2. Leehey P, Naseem M, Every P, et al: Vertebral hemangioma with compression myelopathy: Metrizamide CT demonstration. *J Comput Assist Tomogr* 1985;9:985–986.
3. Lichtenstein L. *Bone Tumors*, ed 4. St Louis, CV Mosby, 1972.
4. Murray RO, Jacobson HG: *The Radiology of Skeletal Disorders: Exercises in Diagnosis*. Edinburgh, Churchill Livingstone, 1977.
5. Schmorl G, Junghanns H: *The Human Spine in Health and Disease*, ed 2. New York, Grune & Stratton, 1971.
6. Unni KK, Ivins JC, Beabout JW, et al: Hemangioma, hemangiopericytoma, and hemangioendothelioma (angiosarcoma) of bone. *Cancer* 1971;27:1403–1414.
7. Yu R, Brunner DR, Rao KCVG: Role of computed tomography in symptomatic vertebral hemangiomas. *CT* 1984;8:311–315.

CASE 30

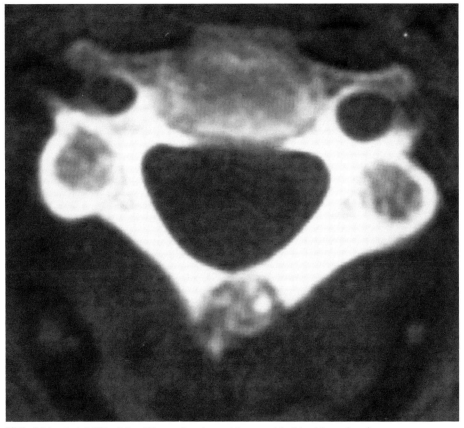

FIG. 30A. Axial CT of the C4 vertebra in an 11-year-old girl with neck pain.

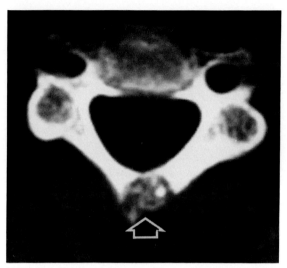

FIG. 30A-1. Osteoblastoma. There is an osteolytic lesion involving the junction of the spinous process and both laminae (*arrow*). Ossifications are noted within the matrix.

Osteoblastoma

There is a destructive lesion with matrix ossification involving the spinous process and adjacent laminae (Fig. 30A-1). This represents an osteoblastoma, a rare, benign primary bone tumor comprising fewer than 1% of all primary tumors of bone.[9] Approximately 35% of osteoblastomas arise in the spine or the sacrum,[6] occurring in decreasing frequency in the lumbar, thoracic, and cervical spine and the sacrum. Osteoblastoma usually develops within the first three decades of life.

Most cases of osteoblastoma are detected by conventional radiography. CT can be helpful in determining the exact anatomic location and extent of the tumor and its relationship to surrounding structures.[2,7,8,11] The CT findings of osteoblastoma are those of an osteolytic, frequently expansile lesion usually involving the posterior elements (pedicle, lamina, articular process) alone or to a lesser extent in combination with the vertebral body. Isolated involvement of the vertebral body is uncommon.[9] The margin is usually sharply defined and sclerotic, although aggressive margins simulating malignancy may be present. Osteoblastomas are greater than 1 cm and usually larger than 2 cm in diameter when initially discovered.[7,9] An ossified matrix is identified by conventional radiography in almost 50% of cases;[9] however, it is more readily detected by CT because of its superior resolution. Ossification of tumor in the spinal canal can be identified.[1]

Approximately 25% to 50% of patients with osteoblastoma present with neurologic signs due to expansion of tumor into the spinal canal.[7,9] In this clinical setting conventional radiographs may demonstrate the lesion. CTM then follows as an excellent imaging modality not only to delineate the location and extent of tumor in the spine and paraspinal compartments, but also to determine the degree of spinal cord or cauda equina compression.[12]

Osteoid osteoma is a lesion that is histopathologically similar to osteoblastoma. Although only 10% of osteoid osteomas occur in the spine, they are more common than osteoblastomas of the spine.[6] Osteoid osteoma almost always involves the posterior elements—particularly the lamina, articular process, or pedicle—whereas osteoid osteoma isolated to the vertebral body occurs in fewer than 10% of cases.[7] Approximately 60% of spinal osteoid osteomas develop in the lumbar region.[7]

Osteoid osteoma may be difficult to detect initially by conventional radiography because of overlying bony structures and the small size of the lesion.[3,7] When visible, it usually appears as a radiolucent lesion (nidus) measuring 1 cm or less and surrounded by exuberant sclerosis.[7] Ossification, which may be present within the nidus, is often difficult to detect. Sometimes the lesion appears entirely osteoblastic if the nidus is not observed (Fig. 30B). Scoliosis is frequent, with the lesion located on the concave side of the scoliotic curve. Radionuclide bone scanning is almost always positive in cases of osteoid osteoma[3] and may be the first modality to detect this lesion. However, the radionuclide bone scan, although highly sensitive, lacks specificity.

The CT appearance of osteoid osteoma is characteristic, with a radiolucent nidus usually surrounded by intense reactive sclerosis[3,6] (Fig. 30C). In some

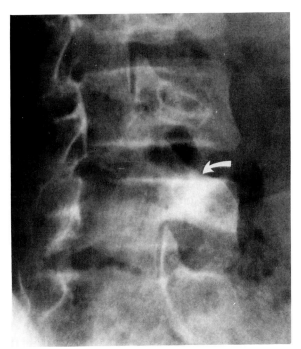

FIG. 30B. Osteoid osteoma. Conventional radiograph in the left posterior oblique projection. There is osteosclerosis of the left pedicle of L5 (*arrow*). No radiolucent nidus is identified.

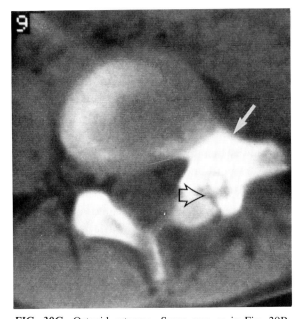

FIG. 30C. Osteoid osteoma. Same case as in Fig. 30B. Axial CT examination demonstrates a radiolucent nidus (*short arrow*) that was not visible on the conventional radiographic examination. Central ossification of the nidus is identified. Reactive sclerosis is present in the pedicle (*long arrow*) and around the lesion. (Figs. 30B and 30C courtesy of Jay Mall, M.D., San Francisco, Calif.)

cases CT may be the only imaging modality to demonstrate the nidus.[3] Ossification that may be present within the nidus is more readily detected by CT than by conventional radiography.[7] Osteoid osteoma remains localized and does not progressively destroy or expand surrounding bone or extend into the paravertebral or intraspinal compartments as might osteoblastoma. Mild enlargement of posterior elements may develop due to cortical thickening from periosteal bone formation rather than marrow expansion.[6] CT can detect the precise anatomic location of the nidus; this aids the surgeon in determining the best surgical approach and obviates unnecessary resection, which otherwise might lead to compromise of spine stability.[3,6,7,10,11,16] When osteoid osteoma involves the pedicle, detection of the nidus differentiates osteoid osteoma from other causes of an osteosclerotic pedicle[16] such as osteoblastic metastasis, Paget's disease, and pedicle sclerosis secondary to a contralateral pars defect or absent pedicle. Occasionally, the nidus of osteoid osteoma is not visible on CT because of partial volume averaging.

Aneurysmal bone cyst is a tumorlike lesion that most often occurs during the first three decades[5] (Fig. 30D). Approximately 15% to 20% of aneurysmal bone cysts develop in the spine,[5,13,15] with a predilection for the cervical and thoracic regions.[13] It usually begins eccentrically in the vertebral body or in the lamina or pedicle, and then may extend into adjacent bony structures and soft tissues.[13] Thus, when discovered, aneurysmal bone cysts may involve any of the posterior elements alone or in combination, or involve the posterior elements along with the vertebral body. Aneurysmal bone cyst is an osteolytic lesion that may become highly expansile. A thin rim of bone that surrounds the lesion may not be detected with conventional radiography. Sometimes the margin of the lesion may be poorly defined and appear aggressive. Aneurysmal bone cysts may extend to involve an adjacent vertebral level, a feature that is unusual for benign and most malignant tumors of the spine.[13] CT is helpful in evaluating not only the osseous extent of aneurysmal bone cyst but also its extent into the paravertebral and intraspinal compartments.[14,15] It can delineate the faintly calcified thin rim that often eludes detection on the conventional radiograph.[4] Patients with aneurysmal bone cyst may present with signs of spinal cord or cauda equina compression. In this clinical setting, CTM can aid in determining the degree of neural compression. In older patients, slow-growing metastases and myeloma can also appear as osteolytic expansile lesions involving the posterior elements alone or in combination with the vertebral body.

Case 30

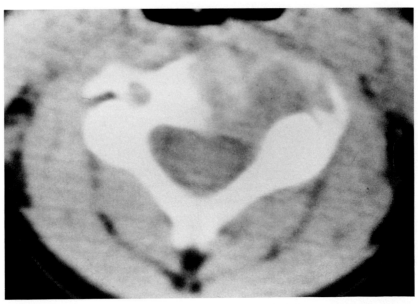

FIG. 30D. Aneurysmal bone cyst. Axial CT at C4 in a 5-year-old child. A large osteolytic lesion is destroying the left half of the vertebral body, the adjacent pedicle, and the bone surrounding the foramen transversarium. No calcification is visible within the matrix. Eosinophilic granuloma could have a similar CT appearance.

References

1. Amacher AL, Eltomey A: Spinal osteoblastoma in children and adolescents. *Childs Nerv Syst* 1985;1:29–32.
2. Epstein N, Benjamin V, Pinto R, et al: Benign osteoblastoma of the thoracic vertebra. *J Neurosurg* 1980;53:710–713.
3. Gamba JL, Martinez S, Apple J, et al: Computed tomography of axial skeletal osteoid osteomas. *AJR* 1984;142:769–772.
4. Haney P, Gellad F, Swartz J: Aneurysmal bone cyst of the spine. Computed tomographic appearance. *J Comput Tomogr* 1983;7:319–322.
5. Hay MC, Paterson D, Taylor TKF: Aneurysmal bone cysts of the spine. *J Bone Joint Surg Br* 1978;60-B:406–411.
6. Jackson RP, Reckling FW, Mantz FA: Osteoid osteoma and osteoblastoma: Similar histologic lesions with different natural histories. *Clin Orthop* 1977;128:303–313.
7. Janin Y, Epstein JA, Carras R, et al: Osteoid osteomas and osteoblastomas of the spine. *Neurosurgery* 1981;8:31–38.
8. Kirwan EO'G, Hutton PAN, Pozo JL, et al: Osteoid osteoma and benign osteoblastoma of the spine. *J Bone Joint Surg Br* 1984;66-B:21–26.
9. McLeod RA, Dahlin DC, Beabout JW: The spectrum of osteoblastoma. *AJR* 1976;126:321–335.
10. Nelson OA, Greer RB III: Localization of osteoid-osteoma of the spine using computerized tomography. *J Bone Joint Surg Am* 1983;65-A:263-264.
11. Omojola MF, Cockshott WP, Beatty EG: Osteoid osteoma: An evaluation of diagnostic modalities. *Clin Radiol* 1981;32:199–204.
12. Omojola MF, Fox AJ, Viñuela FV: Computed tomographic metrizamide myelography in the evaluation of thoracic spinal osteoblastoma. *AJNR* 1982;3:670–673.
13. Tillman BP, Dahlin DC, Lipscomb PR, et al: Aneurysmal bone cyst: An analysis of ninety-five cases. *Mayo Clin Proc* 1968;43:478–495.
14. Volikas Z, Singounas E, Saridakes G, et al: Aneurysmal bone cyst of the spine: Report of a case. *Acta Radiol Diagn* 1982;23:643–646.
15. Wang AM, Lipson SJ, Haykal HA, et al: Computed tomography of aneurysmal bone cyst of the L1 vertebral body: Case report. *J Comput Assist Tomogr* 1984;8:1186–1189.
16. Wedge JH, Tchang S, MacFadyen DJ: Computed tomography in localization of spinal osteoid osteoma. *Spine* 1981;6:423–427.

CASE 31

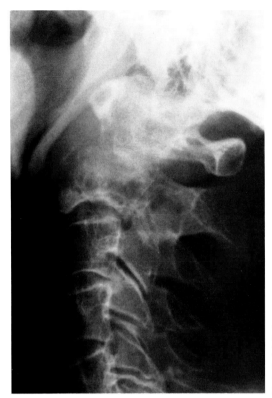

FIG. 31A. Laterial radiograph of the cervical spine. This 57-year-old male had an 18-month history of neck pain and a 4-month history of paresthesias of the hands. He had decreased skilled motor function and difficulty walking.

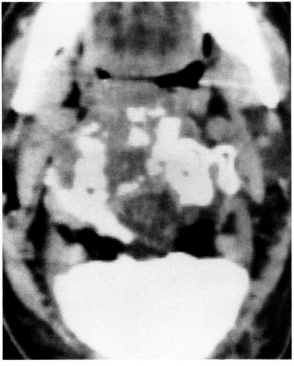

FIG. 31B. Axial CT at C2 obtained after a myelogram had demonstrated block at C2-C3.

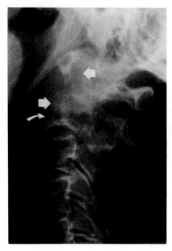 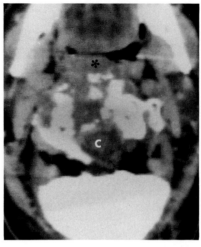

FIG. 31A-1. Chordoma of C2. There is destruction of the C2 vertebral body and odontoid process (*straight arrows*). A pathologic fracture of the C2 vertebral body is present (*curved arrow*).

FIG. 31B-1. Axial CT demonstrates extensive destruction of the C2 vertebral body with calcification or bony fragments noted. Tumor extends anteriorly (*asterisk*), causing compression of the airway, and posteriorly, causing compression of the spinal cord (*C*).

Chordoma of the Spine

There is destruction of the C2 vertebral body and odontoid process with a large anterior paravertebral mass compressing the airway (Figs. 31A-1, 31B-1). Calcification or bony fragmentation is present within the mass, which also extends posteriorly into the spinal canal, compressing the subarachnoid space. This represents a chordoma of C2.

Chordoma is a rare primary bone tumor that arises from notochord rests. A review of several large series finds chordoma most prevalent in the sacrococcygeal region (50%), followed by the clivus (35%) and vertebral column (15%).[1,9] Chordoma occurs more frequently in the cervical and lumbar spine than in the thoracic region.[5,6,9] Pain is the most frequent initial complaint of patients with chordoma of the cervical spine. Occasionally these patients present with dysphagia or with respiratory difficulty caused by soft-tissue extension of the mass into the posterior pharynx.[8] Although slow-growing, chordoma frequently metastasizes.[9]

The CT appearance of chordoma of the spine is that of an osteolytic or mixed osteolytic and osteoblastic lesion of the vertebral body (Fig. 31C). Rarely, chordoma may be entirely osteoblastic, resembling an "ivory" vertebra.[7] Extension into adjacent posterior elements is not common initially but occurs frequently with recurrent disease.[5] Chordoma is usually associated with an anterior or lateral paraspinal mass[5] that is often much larger than might be expected from the amount of bone involvement.[3] CT is more sensitive in detecting calcification than is conventional radiography. Amorphous calcification of the soft-tissue mass is found by CT in 40% of vertebral chordomas[5] and in an even greater percentage of sacral chordomas.[3] Calcification tends to be more exten-

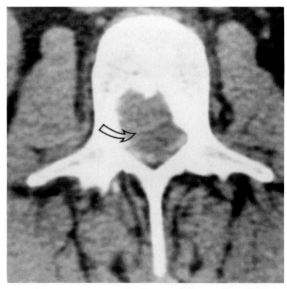

FIG. 31C. Chordoma of the lumbar spine. The osteolytic lesion of the L3 vertebral body extends into the epidural space (*arrow*). No matrix calcification is identified. Metastasis has a similar CT appearance. (Figure courtesy of Nancy Sherwin, M.D., Bridgeton, N.J.)

sive at the periphery of the tumor.[3] The soft-tissue mass may have a sharply defined margin due to a fibrous pseudocapsule.[5] In over 50% of vertebral chordomas there are single or multiple areas of low attenuation within the tumor mass,[5] probably representing myxomatous or gelatinous tissue.[9] Chordoma frequently extends into the epidural space, obliterating epidural fat and appearing as a soft-tissue mass within the spinal canal.[3,5,9] There may be involvement of two contiguous vertebral bodies with invasion of the intervertebral disc; this feature is unusual in other tumors of the spine.[1,2,6] Rarely, a cervical chordoma may enlarge the intervertebral foramen and simulate a neurofibroma.[10]

Although conventional radiography may detect vertebral body destruction and soft-tissue extension, CT has proven to be the most useful modality for demonstrating the full extent of the tumor.[9] CT affords detailed appraisal of soft-tissue extension and impingement on paravertebral structures such as the trachea and esophagus with cervical lesions and the rectum and bladder with sacral tumors. When sacral chordoma predominantly involves the soft tissues, the diagnosis may be established by CT despite a normal radiographic examination.[4] CT is useful in planning the proper surgical approach for spinal chordoma. When radiotherapy is needed, CT best demonstrates the extent of tumor, thus permitting delivery of higher doses of radiotherapy to the tumor while normal tissues are spared.[3,9] CT also aids in the evaluation of tumor extent following therapy. If signs of spinal cord compression are present, CTM can better delineate the relationship of the tumor to the spinal cord or cauda equina.[1,11]

The differential diagnosis of chordoma includes metastasis, chondrosarcoma, and osteosarcoma. Osseous metastases cause vertebral destruction, and although they do not calcify, they may displace islands of bone into the paravertebral and epidural compartments, thus resembling a calcified mass. Chondrosarcoma is a malignant tumor of cartilage matrix that is rare in the spine and sacrum. The CT appearance is usually that of an eccentrically placed osteolytic lesion with extensive punctate calcifications of tumor matrix. Sometimes the calcifications may appear partly amorphous when calcification is extensive or when the tumor is aggressive. Tumor frequently extends into the paravertebral and epidural compartments. Chondrosarcoma may arise de novo or from malignant degeneration of an osteochondroma (Figs. 31D, 31E). Osteosarcoma is a highly malignant bone tumor that rarely involves the spine. It produces radiodense, amorphous tumor bone that often extends into the paravertebral soft tissues (Fig. 31F). It may arise de novo or secondary to radiation, or rarely from sarcomatous degeneration in Paget's disease.

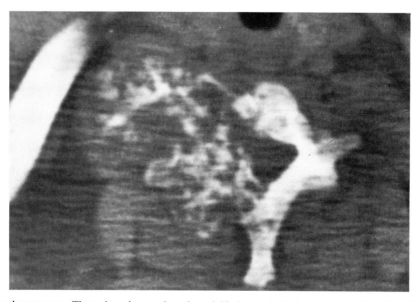

FIG. 31D. Chondrosarcoma. There is a large, densely calcified osteolytic lesion destroying the right half of the C7 vertebral body, pedicle, and lamina. The punctate nature of the calcifications is characteristic of cartilage matrix. The mass extends deeply into the paravertebral soft tissues and intraspinal compartment. (Figure courtesy of George Teplick, M.D., Philadelphia, Pa.).

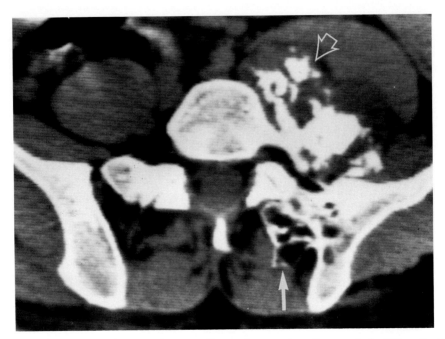

FIG. 31E. Chondrosarcoma occurring secondary to malignant degeneration of an exostosis. This patient has multiple exostoses. There is a benign exostosis derived from the left ilium (*long arrow*). This exostosis has corticated margins. A large mass with irregular, amorphous clumps of calcification is seen adjacent to the left side of L5 (*short arrow*). This proved to be a chondrosarcoma.

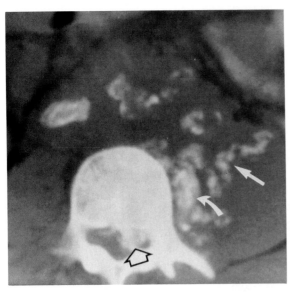

FIG. 31F. Osteosarcoma. There is an osteosclerotic lesion involving the left half of the L4 vertebral body and pedicle. The tumor extends far anteriorly and laterally into the paravertebral soft tissues. Ossifications of the matrix are both punctate (*straight closed arrow*), resembling cartilage matrix, and amorphous (*curved arrow*), typical of osteosarcoma. Tumor also extends into the epidural space (*open arrow*). (Figure courtesy of Daniel Vanel, M.D., Villejuif, France.)

References

1. Firooznia H, Golimbu C, Rafii M, et al: Computed tomography of spinal chordomas. *CT* 1986;10:45-50.
2. Firooznia H, Pinto RS, Lin JP, et al: Chordoma: Radiologic evaluation of 20 cases. *AJR* 1976;127:797–805.
3. Krol G, Sundaresan N, Deck M: Computed tomography of axial chordomas. *J Comput Assist Tomogr* 1983;7:286–289.
4. Lukens JA, McLeod RA, Sim FH: Computed tomographic evaluation of primary osseous malignant neoplasms. *AJR* 1982;139:45–48.
5. Meyer JE, Lepke RA, Lindfors KK, et al: Chordomas: Their CT appearance in the cervical, thoracic and lumbar spine. *Radiology* 1984;153:693–696.
6. Murali R, Rovit RL, Benjamin MV: Chordoma of the cervical spine. *Neurosurgery* 1981;9:253–256.
7. Schwarz SS, Fisher WS III, Pulliam MW, et al: Thoracic chordoma in a patient with paraparesis and ivory vertebral body. *Neurosurgery* 1985;16:100–102.
8. Shallat RF, Taekman MS, Nagle RC: Unusual presentation of cervical chordoma with long-term survival. *J Neurosurg* 1982;57:716–718.
9. Sundaresan N, Galicich JH, Chu FCH, et al: Spinal chordomas. *J Neurosurg* 1979;50:312–319.
10. Wang A-M, Joachim CL, Shillito J Jr, et al: Cervical chordoma presenting with intervertebral foramen enlargement mimicking neurofibroma: CT findings. *J Comput Assist Tomogr* 1984;8:529–532.
11. Zito JL, Davis KR: The role of computed metrizamide myelography in evaluation of extradural extension from vertebral chordoma. *CT* 1980;4:38–42.

CASE 32

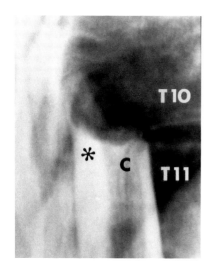

FIG. 32A. Lateral radiograph of the lower thoracic spine from a thoracic myelogram performed with water-soluble contrast introduced into the subarachnoid space at the lumbar level. This 46-year-old female had a history of progressive disturbance of gait and leg weakness.

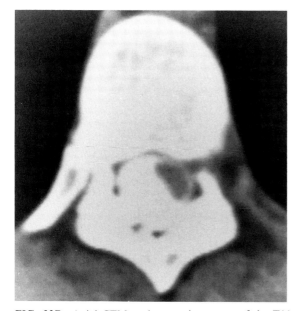

FIG. 32B. Axial CTM at the superior aspect of the T11 vertebra obtained several hours after myelogram.

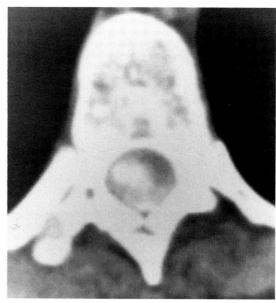

FIG. 32C. Axial CTM scan at the level of T10, 2 cm cephalad to Fig. 32B.

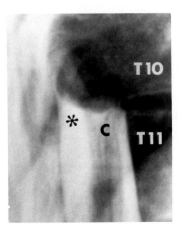 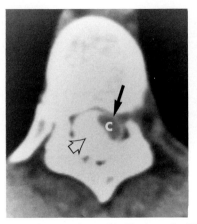 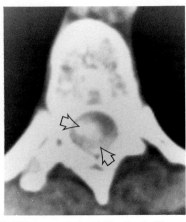

FIG. 32A-1. Meningioma. There is a complete myelographic block at the T10-T11 intervertebral disc level. There is widening of the subarachnoid space posteriorly (*asterisk*), displacement of the spinal cord anteriorly (*C*), and narrowing of the subarachnoid space anteriorly. This appearance is that of an intradural extramedullary lesion located posteriorly.

FIG. 32B-1. CTM just below the level of myelographic block reveals widening of the subarachnoid space posteriorly and on the right (*open arrow*), thinning of the subarachnoid space anteriorly on the left (*closed arrow*), and displacement of the spinal cord (*C*). CTM demonstrates the same typical findings of an intradural extramedullary lesion as are seen with conventional myelography.

FIG. 32C-1. Above the level of the block a mass with amorphous calcification is identified filling the posterior and right side of the canal (*arrows*). This mass measured 2.5 × 1.0 cm and proved to be a psammomatous meningioma. CTM was useful in determining its cephalad extent, which could not be determined by the myelogram.

Meningioma

The myelogram performed with intrathecal water-soluble contrast demonstrates a complete block at the T10-T11 level with the typical configuration of an intradural extramedullary lesion (Fig. 32A-1). CTM examination just below the level of the block confirms widening of the subarachnoid space on the right with displacement of the spinal cord anteriorly and to the left (Fig. 32B-1). Above the block at the T10 level is a densely calcified mass—a meningioma (Fig. 32C-1). Despite the presence of a myelographic block, CTM was able to demonstrate the cephalad extension of the tumor.

In adults, meningioma is second only to neurofibroma as the most frequent primary intraspinal tumor. Typically, meningiomas occur in the thoracic spine in middle-aged women. Meningiomas are less frequent in the cervical spine and are rare in the lumbar region.[7] Predilection for the thoracic spine is not evident in men. Meningiomas are rare in children.[1] Thoracic meningiomas are usually located posteriorly, whereas cervical meningiomas usually develop anterior to the spinal cord.[2] Most meningiomas are purely intradural extramedullary in location, with only about 7% involving both the intradural and extradural compartments.[2] Actually meningioma is the most frequent tumor confined to the intradural extramedullary location, since one third of neurofibromas involve the extradural compartment. Meningioma is also the intraspinal tumor that most frequently calcifies.

When an intraspinal lesion is suspected clinically, conventional radiography is usually followed by conventional myelography[6] or magnetic resonance imaging in an attempt to localize the lesion. Conventional myelography can adequately demonstrate the presence of tumor and its compartmental location (i.e., extradural, intradural extramedullary, or intramedullary). Typically, intradural extramedullary tumors displace the spinal cord to the opposite side, cause widening of the subarachnoid space on the side of the tumor, and cause narrowing or obliteration of the contralateral subarachnoid space (Figs. 32A-1, 32D, 32E). In general, CTM helps one further evaluate the full extent of an intraspinal tumor and determine its relationship to the spinal cord.[5,6] CTM is particularly useful when there is a "complete" myelographic block. Small amounts of contrast that pass beyond the block may not be seen with conventional myelography but may be visualized on CTM examination.[6] Information gained from the CT study may be useful in planning the surgical approach.[3] For example, a densely

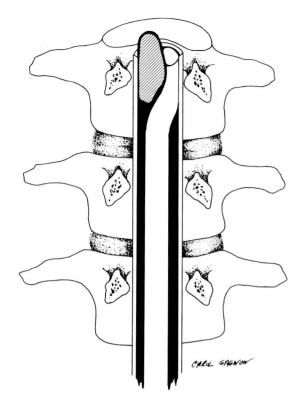

FIG. 32D. An intradural extramedullary tumor (*shaded area*) and its relationship to the subarachnoid space (*black area*) and spinal cord in the coronal plane. This type of tumor causes displacement of the spinal cord, widening of the ipsilateral subarachnoid space, and narrowing of the contralateral subarachnoid space.

calcified or ossified meningioma may occasionally be adherent to the cord and more difficult to remove surgically. This is especially true if the tumor is located anteriorly, where it is less accessible to the surgeon.[2] Extradural extension of meningioma can also be detected by CT and suggests a more invasive tumor and a more ominous course.[2] These tumors may invade bone and resemble metastatic disease.

The differential diagnosis of an intradural extramedullary tumor is usually between neurofibroma and meningioma. Several characteristic CT findings may be useful in differentiating these two tumors. Meningiomas, for example, may contain globular or punctate calcifications that are more readily visualized by CT than by conventional radiography[4] (Fig. 32F). Reactive sclerosis of the adjacent vertebral body may be seen in association with meningioma. Neurofibromas, on the other hand, typically do not calcify and are not associated with vertebral sclerosis. However, other osseous abnormalities are more frequent with neurofibroma than with meningioma and include widening of the neural foramen and vertebral body erosion. A "dumbbell" tumor (i.e., an intradural tumor that extends into and beyond the neural foramen) is more likely to represent neurofibroma than meningioma (Fig. 32G). Both meningiomas and neurofibromas may enhance after intravenous contrast; however, contrast enhancement of meningioma is not readily apparent if the tumor is densely calcified.

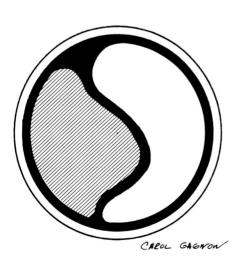

FIG. 32E. Intradural extramedullary tumor (*shaded area*) and its relationship to the subarachnoid space and spinal cord in the axial plane. The effects of an intradural extramedullary tumor on the spinal cord and subarachnoid space are similar to that seen with axial CTM.

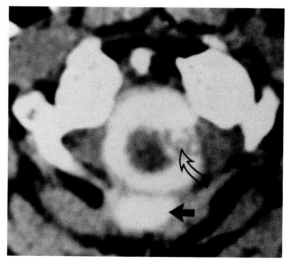

FIG. 32F. Recurrent meningioma. This 55-year-old female had previous surgery for a meningioma at the craniocervical junction. Recurrent symptoms led to CTM examination. Punctate calcifications within an intradural extramedullary mass indicate recurrrent tumor (*curved arrow*). There is compression of the spinal cord. Metrizamide contrast fills the subarachnoid space and has entered a postoperative pseudomeningocele (*straight arrow*).

Case 32

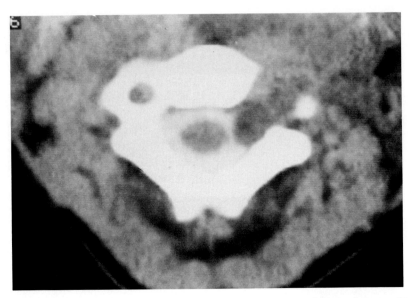

FIG. 32G. Neurofibroma. CTM examination at C2. This 9-year-old girl has a dumbell-shaped neurofibroma filling and widening the neural foramen on the left. There is also compression of the contrast-filled subarachnoid space. The location and configuration of the lesion and the lack of calcification are more typical of a neurofibroma than of meningioma. (Figure courtesy of Spencer Borden IV, M.D., Philadelphia, Pa.)

References

1. DeSousa AL, Kalsbeck JE, Mealey J Jr, et al: Intraspinal tumors in children. *J Neurosurg* 1979;51:437–445.
2. Levy WJ Jr, Bay J, Dohn D: Spinal cord meningioma. *J Neurosurg* 1982;57:804–812.
3. Memon MY, Schneck L: Ventral spinal tumor: The value of computed tomography in its localization. *Neurosurgery* 1981;8:108–111.
4. Nakagawa H, Huang YP, Malis LI, et al: Computed tomography of intraspinal and paraspinal neoplasms. *J Comput Assist Tomogr* 1977;1:377–390.
5. Resjö IM, Harwood-Nash DC, Fitz CR, et al: CT metrizamide myelography for intraspinal and paraspinal neoplasms in infants and children. *AJR* 1979;132:367–372.
6. Tadmor R, Cacayorin ED, Kieffer SA: Advantages of supplementary CT in myelography of intraspinal masses. *AJNR* 1983;4:618–621.
7. Wood JB, Wolpert SM: Lumbosacral meningioma. *AJNR* 1985;6:450–451.

CASE 33

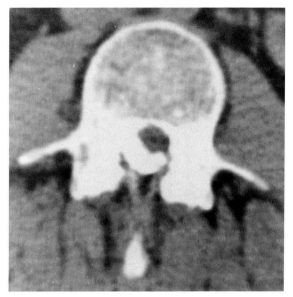

FIG. 33A. Axial CTM at L4 in a 28-year-old woman.

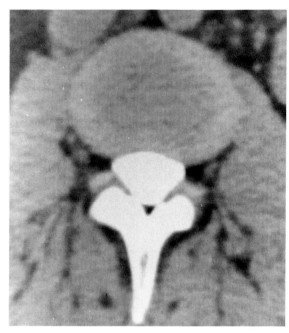

FIG. 33B. Axial CTM at the L3–L4 disc level, 12 mm cephalad to Fig. 33A.

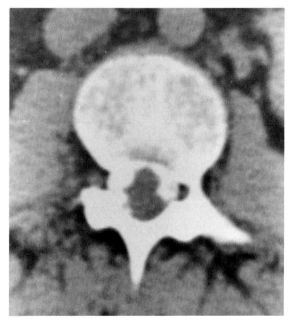

FIG. 33C. Axial CTM at L3, 12 mm cephalad to Fig. 33B.

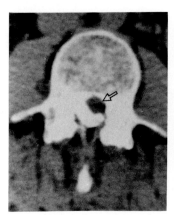 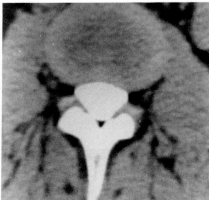 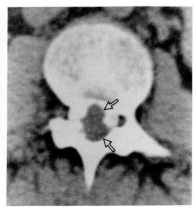

FIG. 33A-1. Multiple intradural schwannomas. At L4 there is an intradural mass (*arrow*).

FIG. 33B-1. At the L3–L4 disc level there is no abnormality. The thecal sac is well distended with intrathecal contrast.

FIG. 33C-1. At L3 a lobulated mass partially fills the thecal sac (*arrows*). This represents a second tumor, separated from the mass at L4 by the normal thecal sac shown in Fig. 33B-1.

Multiple Intradural Tumors: Schwannomas

The CTM examination demonstrates multiple intradural tumors in the lumbar region (Figs. 33A-1–33C-1). Extradural extension of the tumors is not present. Sagittal reconstruction of the axial CTM images demonstrates a complete block at the L3 level (Fig. 33D). At surgery four intradural schwannomas were found, the largest occurring at L3, the site of block. This patient had no stigmata of neurofibromatosis.

A schwannoma is a tumor of nerve sheath origin that occurs along nerve roots or peripheral nerves.[2] These tumors are usually solitary but may be multiple and sometimes occur in patients with neurofibromatosis.[2,5] Most schwannomas occur in the intradural compartment. Neurofibroma is a tumor of nerve root or peripheral nerve origin, usually occurring at multiple sites in association with neurofibromatosis.[2]

In addition to schwannomas and neurofibromas, the differential diagnosis of multiple intradural tumors (either extramedullary or intramedullary) includes metastases, meningiomas, ependymomas, and hemangioblastomas. Isolated intradural metastases are not common and may develop from cerebrospinal fluid seeding of primary intracranial tumors such as me-

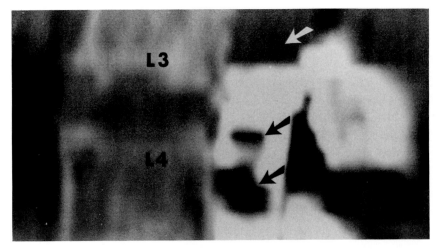

FIG. 33D. Multiple intradural schwannomas. Same patient as in Fig. 33A-1 with sagittal reconstruction of axial CTM images. Intradural tumors are demonstrated at the L3 and L4 levels (*arrows*) with a block to the flow of contrast present at L3.

dulloblastoma, ependymoma, or other gliomas,[6] or as tertiary deposits from cerebral metastasis,[4] usually carcinoma of the breast or lung. Intradural metastases rarely develop from hematogenous spread.

Multiple meningiomas are rare and are usually associated with schwannomas or gliomas in patients with neurofibromatosis. They may also develop following radiation therapy,[1,3] with the spine having been included in the radiation portal. When multiple meningiomas occur, both intracranial and intraspinal lesions may be present.[1] The meningiomas may be so numerous that it may be impossible to differentiate between multiple meningiomas and meningiomatosis.[1] Although multiple spinal ependymomas may occur from metastasis of an intracranial tumor, they may also arise as primary tumors usually associated with neurofibromatosis.[5] Hemangioblastoma may involve the spinal cord or filum terminale and may be multiple, associated with Von Hippel-Lindau disease.[5] These tumors are highly vascular and demonstrate marked intravenous contrast enhancement on CT examination.

Myelography is helpful in the evaluation of possible multiple intraspinal tumors because of its ability to evaluate all levels of the spinal canal. CTM can be used to define the myelographic abnormalities further and to evaluate for possible extradural or extraspinal extension of tumor.

References

1. Holliday PO III, Davis C Jr, Angelo J: Multiple meningiomas of the cervical spinal cord associated with Klippel-Feil malformation and atlantooccipital assimilation. *Neurosurgery* 1984; 14:353–357.
2. Lott IT, Richardson EP Jr: Neuropathological findings and the biology of neurofibromatosis. *Adv Neurol* 1981;29:23–32.
3. Patronas NJ, Brown F, Duda EE: Multiple meningiomas in the spinal canal. *Surg Neurol* 1980;13:78–80.
4. Perrin RG, Livingston KE, Aarabi B: Intradural extramedullary spinal metastasis. *J Neurosurg* 1982;56:835–837.
5. Russell DS, Rubinstein LJ: *Pathology of Tumours of the Nervous System*, ed 4. Baltimore: Williams & Wilkins, 1977.
6. Stanley P, Senac MO Jr, Segall HD: Intraspinal seeding from intracranial tumors in children. *AJR* 1985;144:157–161.

CASE 34

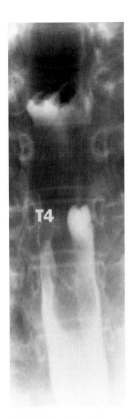

FIG. 34A. AP radiograph of the upper thoracic spine from a myelogram performed with water-soluble contrast introduced at the lumbar level. This 25-year-old had weakness of the arms and legs, inability to walk, and urinary tract difficulties. Surgery had been performed 2 years earlier under similar circumstances.

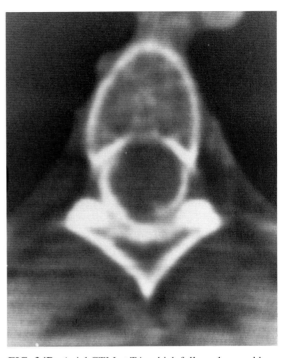

FIG. 34B. Axial CTM at T4, which followed several hours after the conventional myelogram.

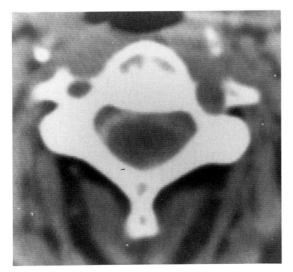

FIG. 34C. Axial CTM at C4.

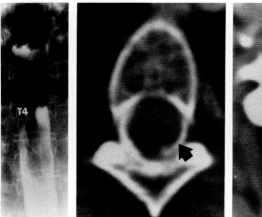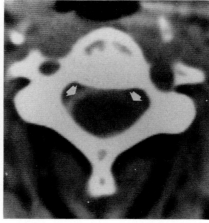

FIG. 34A-1. Intramedullary tumor (ependymoma). There is widening of the spinal cord at T5 and almost complete myelographic block at T4. This patient had had an ependymoma 2 years earlier, and the diagnosis of recurrent tumor was made on the basis of this myelographic study. The cephalad extent of the tumor could not be determined on this examination.

FIG. 34B-1. CTM at T4 demonstrates tumor enlarging the spinal cord and encroaching on the thin, contrast-filled subarachnoid space (*arrow*).

FIG. 34C-1. Despite the myelographic block, some contrast can be seen in the subarachnoid space above the block. On this scan at C4, intramedullary tumor can again be seen widening the cord and compressing the subarachnoid space (*arrows*). The extent of tumor can thus be determined by CTM.

Intramedullary Tumor

This patient had had previous surgery for an intramedullary ependymoma and presented with symptoms of recurrent tumor. A myelogram performed with water-soluble contrast introduced from a lumbar level revealed a block at T4 (Fig. 34A-1). CTM performed several hours later demonstrated an enlarged cord from C4 to T5 outlined by intrathecal contrast in a narrow, compressed subarachnoid space (Figs. 34B-1, 34C-1). There was no evidence of opacification of a syrinx. The diagnosis of recurrent intramedullary ependymoma was established, and the extent of the tumor was determined prior to further surgical intervention.

Gliomas (ependymomas and astrocytomas) comprise the majority of intramedullary tumors. The most frequent intramedullary tumor is ependymoma. Astrocytoma is the second most common and in some series is the most frequent glioma in children.[3] These tumors may be focal or quite extensive, involving the entire cord (holocord);[1,6] this is a common finding with astrocytomas in children.[6] Less frequent intramedullary tumors include hemangioblastoma, metastasis, and lipoma.

Symptoms associated with a spinal cord tumor depend on the spinal level involved. Cervical tumors cause upper extremity weakness and cervical pain. Thoracic tumors present with mild progressive paraparesis and pain over the thoracic spine. These patients may have disturbance of the bowel and bladder sphincters.[6] Some patients may present with gait disturbance, whereas others present with scoliosis without neurologic signs.[2] Idiopathic scoliosis in a young person is painless and usually not associated with neurologic deficit. Painful scoliosis should raise the question of an underlying pathologic process that can be further evaluated by radionuclide bone scanning, myelography, and CTM.[2] Conventional radiography may reveal clues to the presence of an intramedullary tumor. The most frequent radiographic changes are widening of the interpedicular distance over several segments; thinning, flattening, or erosion of the pedicles; and increase in the sagittal diameter of the spinal canal. In the cervical spine evaluation of the interpedicular distance is not reliable. Thus, an increase in the sagittal diameter of the cervical canal is more meaningful than an apparent increase in the interpedicular distance.[1]

CT without contrast is not adequate for evaluation of an intramedullary tumor. When a spinal cord tumor is suspected clinically, myelography with water-

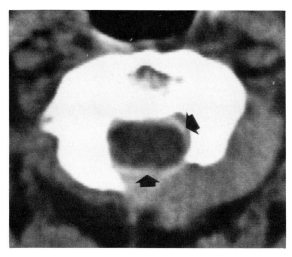

FIG. 34D. Astrocytoma. Axial CTM demonstrates a recurrent astrocytoma that is widening the spinal cord. The contrast-filled subarachnoid space (*arrows*) is thin as a result of the intramedullary mass. A laminectomy defect is noted. This 33-year-old had recurrent tumor extending from C1 to T1, which caused weakness in all extremities, difficulty with locomotion, and burning dysesthesia.

soluble contrast and CTM are used to determine the location and extent of the lesion, the spinal compartment involved, and the presence of a coexistent syrinx. An intramedullary tumor infiltrates and swells the spinal cord, thus thinning the surrounding contrast-filled subarachnoid space (Figs. 34D–34F). When a complete myelographic block is found, CTM can often detect contrast beyond the block, permitting evaluation of the extent of the lesion. Delayed CTM performed 6 to 24 hours after the introduction of intrathecal contrast may demonstrate a syrinx cavity within an intramedullary tumor.[8] At autopsy, approximately 30% of intramedullary tumors are associated with coexistent syrinx.[8] Holocord astrocytomas in children often have cystic cavities rostral and caudal to the solid component of the tumor.[6] Some authors have recommended surgical removal of the solid tumor and drainage of the associated cysts.[6] CTM has been used as an adjunct to myelography in these cases, and the demonstration of contrast within the cystic cavity may sometimes help localize the cystic component of these tumors. Solitary intramedullary metastasis, which is quite rare, has also been reported to demonstrate central contrast filling during CTM or delayed CTM studies.[10]

Although myelography and CTM are important modalities for the evaluation of intramedullary tumors, some authors have used intravenous contrast-enhanced CT to demonstrate these lesions.[5,7,9] The degree of enhancement varies with different tumors and even among patients with the same type of tumor.

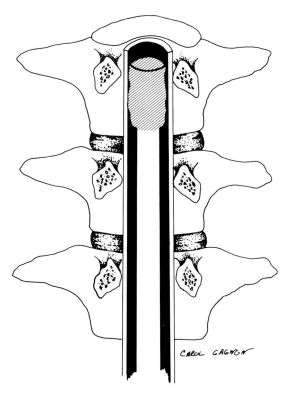

FIG. 34E. Relationship of an intramedullary tumor to the thecal sac in the coronal plane. *Shaded area* indicates the intramedullary mass. Notice the symmetric thinning of the subarachnoid space (black *area*).

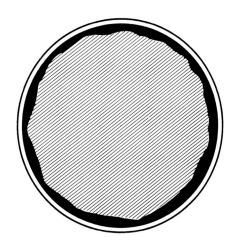

FIG. 34F. Relationship of an intramedullary tumor to the thecal sac in the axial plane. This representation highlights the features of an intramedullary tumor similar to that seen with axial CTM. The intramedullary mass (*shaded area*) compresses the subarachnoid space (*black area*).

In general, however, ependymomas and astrocytomas have mild to marked enhancement, whereas hemangioblastomas typically have marked enhancement due to their highly vascular nature.[9] It is not yet possible to differentiate astrocytoma from ependymoma on the basis of contrast enhancement characteristics. However, intravenous contrast enhancement can help differentiate the solid and cystic components of the tumor. Contrast enhancement may be solid, patchy, or ringlike. The solid component of the intramedullary tumor may appear as an enhancing nodule or mass within the cord[7,9] (Fig. 34G). In some cases several rounded, low-density regions have been demonstrated within a spinal cord tumor after intravenous contrast enhancement[5,9] (Fig. 34H). These low-density areas represent multiloculated cysts. In the patient who has

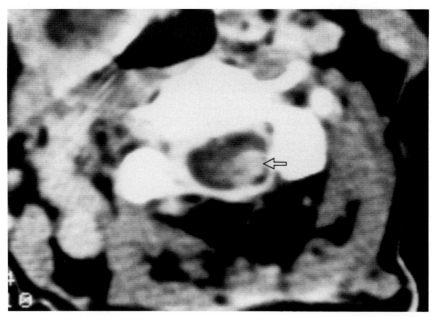

FIG. 34G. Intramedullary metastasis. Axial CT of the cervical spine following intravenous injection of iodinated contrast demonstrates marked enhancement of the tumor (*arrow*). This patient has metastasis to the spinal cord from a papillary carcinoma of the thyroid gland.

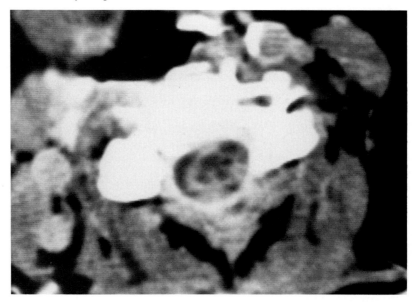

FIG. 34H. Intramedullary metastasis. Same patient as in Fig. 34G. At another cervical level the cystic nature of this tumor could be appreciated. Following injection of intravenous contrast, multiple small, low-density cysts could be seen within the tumor.

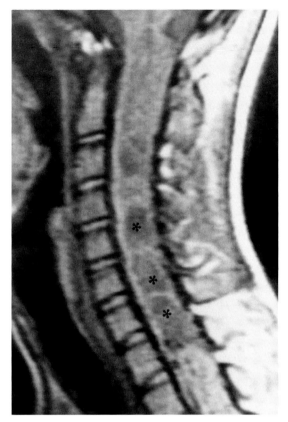

FIG. 34I. T1-weighted magnetic resonance image of the cervical cord in the sagittal plane. There is widening of the entire cord (holocord) by intramedullary tumor. Large cysts (*asterisks*) producing a lower signal are present within the tumor.

undergone surgical removal of an intramedullary tumor, a clinically suspected recurrent tumor can be evaluated by intravenous contrast-enhanced CT.

With magnetic resonance imaging, the location and extent of cord tumor can be identified, and the cystic and solid components can be distinguished[4] (Fig. 34I). The ability to obtain direct sagittal images is particularly useful in the evaluation of intramedullary tumors, which may involve long segments of the spinal cord.

References

1. Banna M, Gryspeerdt GL: Intraspinal tumors in children (excluding dysraphism). *Clin Radiol* 1971;22:17–32.
2. Citron N, Edgar MA, Sheehy J, et al: Intramedullary spinal cord tumors presenting as scoliosis. *J Bone Joint Surg Br* 1984;66-B:513–517.
3. DeSousa AL, Kalsbeck JE, Mealey J Jr, et al: Intraspinal tumors in children. *J Neurosurg* 1979;51:437–445.
4. DiChiro G, Doppman JL, Dwyer AJ, et al: Tumor and arteriovenous malformations of the spinal cord: Assessment using MR. *Radiology* 1985; 156:689–697.
5. Enzmann DR, Murphy-Irwin K, Silverberg GD, et al: Spinal cord tumor imaging with CT and sonography. *AJNR* 1985;6:95–97.
6. Epstein F, Epstein N: Surgical treatment of spinal cord astrocytomas of childhood: A series of 19 patients. *J Neurosurg* 1982;57:685–689.
7. Handel S, Grossman R, Sarwar M: Computed tomography in the diagnosis of spinal cord astrocytoma. *J Comput Assist Tomogr* 1978;2:226–228.
8. Kan S, Fox AJ, Viñuela F, et al: Delayed CT metrizamide enhancement of syringomyelia secondary to tumor. *AJNR* 1983;4:73–78.
9. Lapointe JS, Graeb DA, Nugent RA, et al: Value of intravenous contrast enhancement in the CT evaluation of intraspinal tumors. *AJNR* 1985;6:939–943, *AJR* 1986;146:103–107.
10. Reddy SC, Vijayamohan G, Rao GR: Delayed CT myelography in spinal intramedullary metastasis: Case report. *J Comput Assist Tomogr* 1984;8:1182–1185.

CASE 35

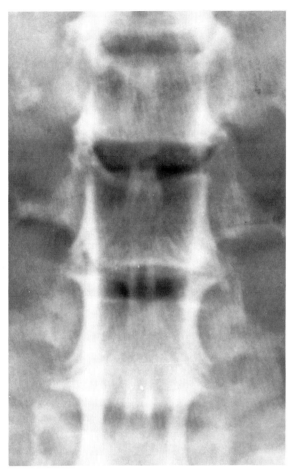

FIG. 35A. PA projection of the lower cervical spine from a cervical myelogram performed with water-soluble contrast. The cross-table lateral radiograph was not diagnostic. This 20-year-old woman had a 2-month history of pain and weakness of the right upper extremity. There were physical findings of cervical radiculopathy and myelopathy.

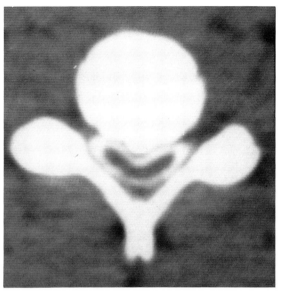

FIG. 35B. Axial CTM just cephalad to the C5-C6 disc space. The CTM examination was performed several hours after the myelogram.

FIG. 35C. Axial CTM at the C5-C6 disc space.

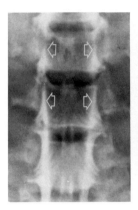 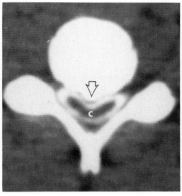 

FIG. 35A-1. There is widening of the spinal cord at C5 and C6 with thinning of the contrast-filled subarachnoid space (*arrows*). This appearance can be caused by an intramedullary lesion or by anterior extradural compression.

FIG. 35B-1. Spondylosis at C5-C6. Metrizamide contrast is well seen in the subarachnoid space. There is a large osteophyte (*arrow*) in the midline compressing the anterior subarachnoid space. The spinal cord (*C*) is compressed and flattened into a heart-shaped configuration.

FIG. 35C-1. Disc herniation and spondylosis at C5-C6. The intrathecal contrast is only faintly visualized anteriorly because of displacement and compression from midline disc herniation (*arrow*) and spondylosis. Disc herniation and spondylosis at C5-C6 along with similar abnormalities at C4-C5 caused the widening of the spinal cord visualized on the myelographic study.

Cervical Spondylosis and Disc Herniation Simulating Myelographic Intramedullary Lesion

The PA projection of the cervical myelogram demonstrates apparent expansion of the spinal cord at the levels of C5 and C6 (Fig. 35A-1). This suggests the presence of an intramedullary lesion; however, very little contrast was seen on the cross-table lateral view, and thus evaluation of this abnormality in a second plane could not be achieved. The patient was therefore further studied by CTM, which demonstrated central, anterior, extradural compression of the subarachnoid space and flattening of the spinal cord into a heart-shaped configuration (Figs. 35B-1, 35C-1). This was caused by cervical spondylosis and disc herniation. The flattened appearance of the spinal cord caused by extradural pathology differs from the uniformly expanded appearance of the cord that occurs secondary to an intramedullary lesion.

Patients with spinal cord compression due to central disc herniation or spondylosis may have symptoms similar to those caused by intramedullary tumor.[2] Extradural pathology may also simulate an intramedullary lesion when the frontal projection of the myelogram is studied.[2,3] The diagnosis may be clarified by the shoot-through lateral radiograph, which demonstrates expansion of the cord when an intramedullary lesion is present and anterior compression when extradural abnormalities are found.[3] However, when the subarachnoid space has been compressed to the extent that little contrast can be seen on the lateral view, CTM is utilized to establish the diagnosis, as in this case. In addition, patients with cervical spondylosis may have reduced flexion and extension with straightening or reversal of the normal cervical lordotic curve. These patients may have inadequate myelographic studies performed with water-soluble contrast even when the contrast is introduced via a C1-C2 puncture.[1] CTM may be invaluable in diagnosing spondylosis and disc herniation in this group of patients with unsatisfactory cervical myelography.[1] In one study, CTM was superior to myelography in the evaluation of patients with central cervical disc herniation and cord compression.[2] CTM clearly delineates the degree and cause of cord compression, rules out the presence of intramedullary tumor, and is diagnostic when a myelographic block is present.

References

1. Fon GT, Sage MR: Computed tomography in cervical disc disease when myelography is unsatisfactory. *Clin Radiol* 1984; 35:47–50.
2. Nakagawa H, Okumura T, Sugiyama T, et al: Discrepancy between metrizamide CT and myelography in diagnosis of cervical disc protrusions. *AJNR* 1983;4:604–606.
3. Shapiro R: *Myelography*, ed 4. Chicago, Year Book Medical Publishers, 1984.

CASE 36

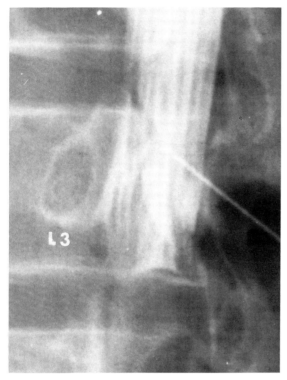

FIG. 36A. Oblique view of a lumbar myelogram performed with water soluble contrast. The intrathecal contrast was introduced at the L2–L3 level. This 36-year-old male had a 2-week history of progressive pain in the right buttock that radiated into the posterior right thigh. He had a history of a similar episode 2 years previously.

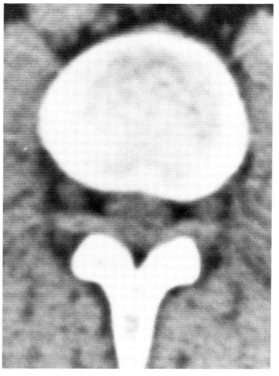

FIG. 36B. CT performed the following day at the L3–L4 level without intrathecal or intravenous contrast.

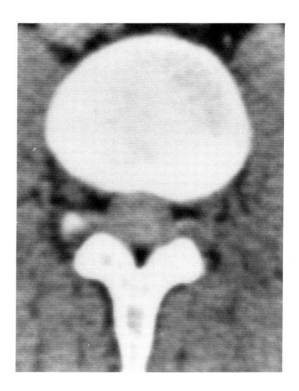

FIG. 36C. CT at same level as in Fig. 36B. This was obtained after bolus injection and drip infusion of intravenous iodinated contrast.

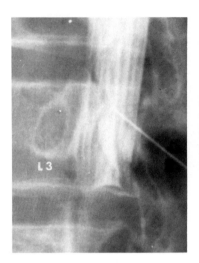 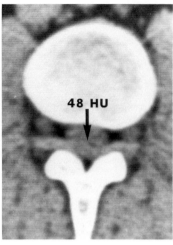 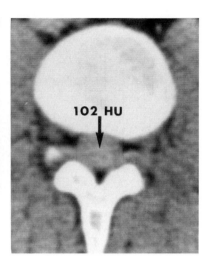

FIG. 36A-1. Ependymoma of the filum terminale. There is a complete block to the flow of contrast at L3–L4.

FIG. 36B-1. The intrathecal contents measure 48 HU. This measurement is slightly above average but is not readily perceived as abnormal.

FIG. 36C-1. After intravenous contrast injection the intrathecal contents measure 102 HU. This contrast enhancement was due to the presence of an ependymoma of the filum terminale.

Ependymoma of the Filum Terminale

This patient's clinical presentation suggested a herniated disc; however, a complete block was demonstrated at the inferior aspect of L3 when a myelogram was performed with water-soluble contrast introduced at L2-L3 (Fig. 36A-1). CTM followed several hours later and demonstrated complete block with no visible contrast caudal to the block. There was no CT evidence of disc herniation at L3-L4 to account for the block, and tumor was suspected. The following day an unenhanced CT scan was performed both above and below the level of the block. The thecal sac at the uninvolved level above the block measured 28 HU, whereas below the level of the block it measured 48 HU (Fig. 36B-1). This density difference is subtle and is not easily perceived visually on the axial scan. Intravenous iodinated contrast was given, and enhancement of an intradural tumor was achieved with measurements of 102 HU recorded (Fig. 36C-1). Sagittal reconstruction demonstrated the tumor (Fig. 36D); however, the lesion was ideally visualized when sagittal reconstruction with the blink mode was used to highlight only the high-density areas (Fig. 36E). Not

 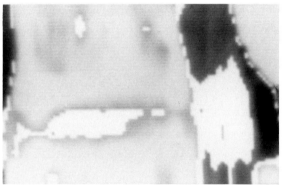

FIG. 36D. Ependymoma of the filum terminale. Same patient as in Fig. 36C-1. Sagittal reconstruction of axial images after intravenous contrast enhancement of ependymoma. The extent and configuration of the tumor are appreciated (*arrows*). A, = anterior; P, = posterior.

FIG. 36E. Ependymoma of the filum terminale. Same sagittal reconstruction as in Fig. 36D studied with highlighting technique, which accentuates tissues within a designated density range. Using this technique, the contrast-enhanced ependymoma is clearly delineated.

all tumors enhance significantly after intravenous contrast injection; however, in this case the use of intravenous contrast clearly demonstrated enhancement and established the presence of an intradural tumor. Sagittal reconstruction allowed precise delineation of the size and extent of the lesion, which proved to be an ependymoma of the filum terminale.

Ependymoma is the most frequent glioma of the spine. Approximately 45% to 75% of spinal ependymomas involve the conus medullaris, filum terminale, or cauda equina.[1,4] These are frequently large, bulky lesions filling the spinal canal and producing myelographic block. Unenhanced CT may show isodensity or slight increased density of the tumor when compared to the normal thecal sac measurements. The lesion may therefore be imperceptible with unenhanced CT technique, especially if the tumor uniformly fills the canal. CTM can sometimes detect intrathecal contrast surrounding tumor even when a "complete block" is present on myelography. However, large, bulky ependymomas of the filum terminale may cause a complete block of contrast even as viewed with CT. In this clinical setting, the extent of the lesion can be myelographically demonstrated by introducing contrast on both sides of the block. However, CT performed after intravenous injection of contrast can sometimes delineate the size, extent, location, and enhancement characteristics of the tumor, thus obviating the need for a second myelographic examination.

Ependymomas of the filum terminale, conus medullaris, or cauda equina may be very extensive and cause scalloping of the vertebral bodies and laminae at multiple levels (Figs. 36F–36H). CT can dem-

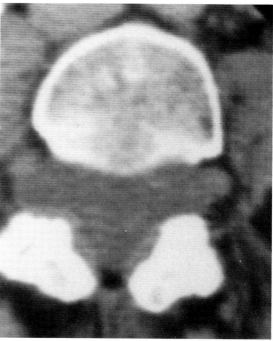

FIG. 36G. Ependymoma. Same patient as in Fig. 36F. Tumor has extended into and has widened both neural foramina.

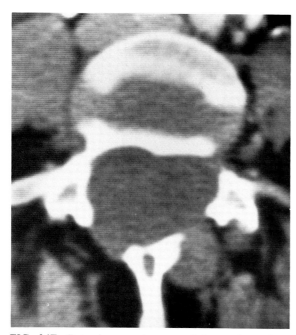

FIG. 36F. Extensive ependymoma of the cauda equina. This CT as well as those shown in Figs. 36G and 36H are representative of the scans obtained throughout the lumbar spine in this patient with a very extensive ependymoma. The bulky tumor causes scalloping of the posterior vertebral body margin and erosion of the laminae. The tumor can be seen extending posteriorly beyond the laminae.

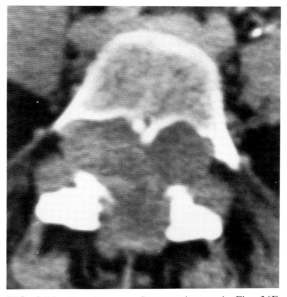

FIG. 36H. Ependymoma. Same patient as in Fig. 36F. The tumor extends laterally and posteriorly beyond the spinal canal. This type of graphic information is important in presurgical planning. Marked scalloping of the posterior vertebral body is demonstrated.

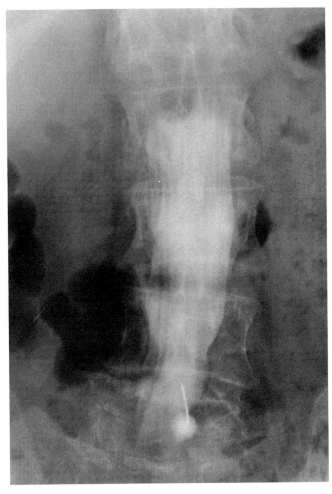

FIG. 36I. Neurofibrosarcoma of the conus medullaris. Myelogram performed with water-soluble contrast introduced at the lower lumbar level demonstrates "complete" myelographic block at the superior aspect of L2. This 73-year-old woman had back and leg pain with generalized weakness of the lower extremities and urinary incontinence.

onstrate tumor extending into the neural foramina and posterior paraspinal musculature. This type of detailed information is very useful in treatment planning. A neural tumor such as a neurofibroma or schwannoma of the cauda equina can have a CT appearance similar to that of an ependymoma and may be very extensive, with osseous erosion and foraminal extension of tumor.[5] Other, less common tumors may involve the conus or the cauda equina. An example of an unusual neurofibrosarcoma is shown and demonstrates the value of CTM (Figs. 36I–36K). In this case a myelographic block was demonstrated at the superior aspect of L2. CTM followed several hours later, and a large intradural mass was demonstrated. The coronal reconstruction view delineates the extent of the lesion.

Clinically, tumors of the cauda equina are a diagnostic challenge. The patients most frequently present with back pain and unilateral sciatica, which may simulate the presentation of a disc herniation.[3] Treatment is often instituted for disc disease, delaying proper diagnosis and therapy. In one series, delay in diagnosis averaged 39 months.[3] Early diagnosis is important because surgical excision of benign tumors of the cauda equina offers an excellent chance of full recovery unless prolonged pressure has led to permanent neurological damage.[3] The tumors that most frequently involve the cauda equina are neurofibroma and ependymoma.

The correct myelographic or CTM diagnosis of tumors of the cauda equina, conus medullaris, or filum terminale requires examination of the abnormal area. A false-negative CT study may be obtained when the examination is limited to the lower lumbar spine. Therefore, when clinical symptoms are atypical or symptoms and signs are not explained by the con-

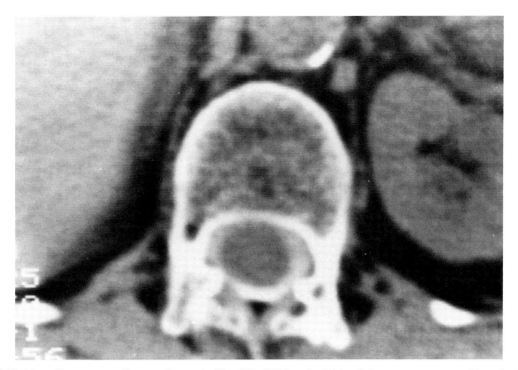

FIG. 36J. Neurofibrosarcoma. Same patient as in Fig. 36I. CTM at the L1 level demonstrates a central intradural mass compressing the contrast-filled subarachnoid space. The mass measures 78 HU. Note that contrast is seen above the level of myelographic block. This assists in delineating the extent of tumor.

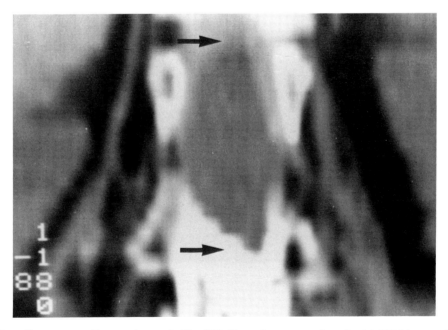

FIG. 36K. Neurofibrosarcoma. Same patient as in Fig. 36I. Coronal reconstruction of axial CTM images. The full extent of the tumor can be appreciated (*arrows*). In this case an unusual neurofibrosarcoma has an appearance similar to the more common ependymoma.

ventional CT study, the region of the conus medullaris can be evaluated by myelography, CTM, or magnetic resonance imaging as indicated.[2]

Reference

1. Barone BM, Elvidge AR: Ependymomas: A clinical survey. *J Neurosurg* 1970;33:428–438.
2. Grogan JP, Daniels DL, Williams AL, et al: The normal conus medullaris: CT criteria for recognition. *Radiology* 1984;151:661–664.
3. Ker NB, Jones CB: Tumours of the cauda equina: The problem of differential diagnosis. *J Bone Joint Surg Br* 1985; 67-B:358-361.
4. Mørk SJ, Løken AC: Ependymoma: A follow-up study of 101 cases. *Cancer* 1977;40:907–915.
5. Osborn RE, DeWitt JD: Giant cauda equina schwannoma: CT appearance. *AJNR* 1985;6:835–836.

CASE 37

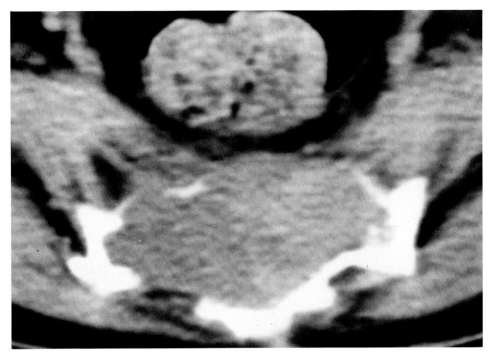

FIG. 37A. CT examination of the sacrum in a 57-year-old male with vague rectal pain.

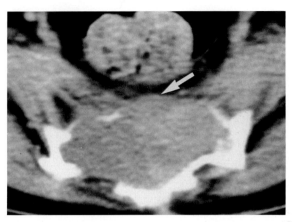

FIG. 37A-1. A large, centrally located osteolytic lesion is present within the sacrum and extends beyond the sacrum anteriorly (*arrow*). The matrix of the lesion is not calcified. This represents recurrent chordoma. Tumor was originally resected 12 years previously.

Chordoma of the Sacrum

A large osteolytic lesion is present in the sacrum and extends anteriorly beyond the bony margin (Fig. 37A-1). No calcifications are identified. The differential diagnosis in a patient this age includes metastasis, myeloma, and chordoma. This tumor represented recurrent chordoma.

The sacrum is a difficult bone to evaluate on conventional radiography because of its curvature and the presence of overlying bowel gas and feces. When sacral lesions are suspected by conventional radiography or other imaging modalities, CT allows excellent evaluation of the bony sacrum, neural foramina, central sacral canal, parasacral soft tissues, and sacroiliac joints.[2,6,9,12,13] Tumors and tumorlike conditions of the sacrum are infrequent, with the exception of metastasis. Some sacral lesions may be differentiated on CT by the pattern of bone destruction, the location of the lesion, and the presence of matrix calcification.

Chordoma is a rare tumor that most frequently involves the sacrum. It most commonly develops during the fourth through the seventh decades. Chordoma usually begins in the midline of the fourth and fifth sacral vertebrae and is typically a progressive, slow-growing tumor that reaches large size.[2] CT demonstrates a mixed osteolytic and osteoblastic pattern, although pure osteolytic lesions may develop[1] (Fig. 37A-1). Calcification is frequently detected with CT and is found more commonly in chordomas of the sacrum than in those of other axial sites[1] (Fig. 37B). Chordoma of the sacrum begins centrally and is often symmetric when discovered. It may involve the sacral foramina and central sacral canal. Extension of tumor into the presacral soft tissues or buttocks is common, and massive soft-tissue involvement may seem out of proportion to the bone destruction.[1] CT can detect involvement of the sacral canal below the level of the thecal sac—an area inaccessible to myelography.[1] Direct coronal CT images are useful for the demonstration of tumor involving the sacral nerve roots.[5] Chordomas may not enhance after the injection of intravenous contrast even when they are histologically vascular lesions.[3] Magnetic resonance imaging (MRI) is another useful modality in the evaluation of sacral chordomas. Compared to CT, MRI has the advantage of direct sagittal imaging, which is helpful in determining the longitudinal extent of the tumor (Fig. 37C). MRI also affords superior contrast between tumor and surrounding soft tissues.[5]

Metastasis is the most frequent tumor of the bony sacrum. It usually develops from hematogenous spread of tumor, most frequently from carcinoma of the breast, lung, or prostate in the adult. However, it may occur by direct extension from carcinoma of the rectum. CT of most metastases demonstrates single or multiple osteolytic lesions (Fig. 37D). Metastasis from carcinoma of the prostate, lymphoma, and previously treated metastases appear osteoblastic (Fig. 37E). Although metastasis frequently occurs in the sacral alae it may be found anywhere in the sacrum.[9] Tumor may be confined to the bony sacrum, often extending to and invading the sacral foramen or central sacral canal.[13] It may destroy surrounding cortex and extend anteriorly into the soft tissues of the pelvis, or may invade posteriorly into the buttocks.[2] Metastasis may cause bone "expansion" or become so extensive that the entire sacrum may be destroyed. When metastasis develops from direct extension from carcinoma of the rectum, the anterior sacral cortex is destroyed and a pelvic mass is usually present. A

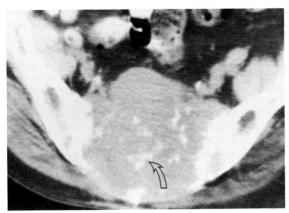

FIG. 37B. Chordoma. This large, centrally located osteolytic lesion demonstrates matrix densities representing either calcification in the tumor or islands of bone displaced by tumor (*arrow*). (Figure courtesy of William Murphy, M.D., St. Louis, Mo.)

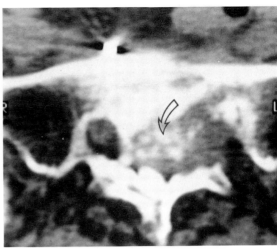

FIG. 37D. Metastasis. Destruction of the sacrum obliterates the left neural foramen (*arrow*). Compare with the normal right sacral foramen in which the sacral nerve and perineural fat are identified. Bony fragments remain within the tumor mass.

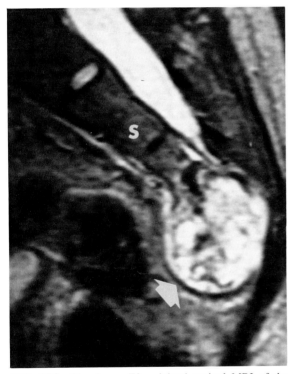

FIG. 37C. Chordoma. T2-weighted sagittal MRI of the sacrum and coccyx of the same patient as in Fig. 37A-1. The tumor is seen as a lobulated mass (*arrow*) in the sacrococcygeal region consisting predominantly of high-intensity signals. The anterior and posterior margins are well defined.

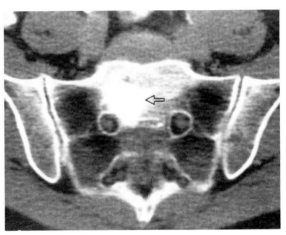

FIG. 37E. Metastasis. An osteoblastic lesion is adjacent to the sacral foramen (*arrow*). This patient has metastatic disease to the sacrum from carcinoma of the prostate gland.

large, solitary, centrally located sacral metastasis may appear similar to a primary bone tumor of the sacrum such as chordoma.[2] Metastasis to bone does not calcify; however, tumor may displace islands of partially destroyed bone, which may on CT appear as calcific densities within the tumor mass (Fig. 37D).

Tumors of nerve (neurofibroma) or nerve sheath (neurilemmoma, ganglioneuroma) that develop in the sacral foramen obliterate perineural fat. With progressive growth, they widen the sacral foramen and eventually destroy surrounding bone (Fig. 37F). The central sacral canal is usually spared with tumors of nerve origin.[13] With more advanced lesions, the zone of destruction is extensive, and tumor may extend into the presacral soft tissues and even into the central sacral canal.

Ewing's sarcoma is a highly malignant bone tumor that usually occurs during the first two decades of life. The sacrum accounts for 60% of the Ewing's sarcomas in the axial skeleton.[14] There is often aggressive osteolytic bone destruction sometimes associated with reactive sclerosis; however, entirely osteoblastic lesions may be encountered. Tumor may invade the presacral region, forming a huge mass (Fig. 37G). Calcification in tumor is extremely unusual in Ewing's sarcoma[11] and when visible may reflect necrosis of tumor or islands of bone displaced by the growing tumor mass. CT is helpful in treatment planning and follow-up evaluation.[10]

Osteosarcoma rarely involves the sacrum. This highly malignant primary bone tumor may arise de novo or secondary to Paget's disease, previous radiation (either internal or external), or rarely other bone lesions. Osteosarcoma is the most common radiation-induced tumor[7] (Figs. 37H, 37I). Approximately 5% of radiation-induced sarcomas occur in the sacrum.[7] The latent period between the radiation and the development of the tumor ranges from 1 to 42 years with a median latent period of 11 years.[4]

Within the axial skeleton, the sacrum is the most frequent site of giant cell tumors,[8] a primary bone tumor that occurs most frequently during the third and fourth decades. Next to chordoma, giant cell tumor is the second most frequent primary tumor of the sacrum.[8] Early, CT may demonstrate an osteolytic lesion with a thin, sclerotic margin.[2] As tumor grows, it may expand the bone symmetrically. Eventually, tumor destroys cortex and extends into the presacral region as a large mass displacing rectum and bladder, or it may extend posteriorly into the buttock.[2]

Osteomyelitis of the sacrum may be associated with infection of the adjacent sacroiliac joint. CT demonstrates osteolytic bone destruction, which is often accompanied by evidence of reparative sclerosis in surrounding bone. Associated soft-tissue mass may be evident. The adjacent sacroiliac joint may be narrow with irregular margins. Osteomyelitis of the sacrum may develop from hematogenous spread without sacroiliitis or develop secondary to a deep decubitus ulcer. In these clinical settings, the sacroiliac joints may be spared.

The CT appearance of Paget's disease depends on its stage. It may appear osteolytic and/or osteoblastic (Figs. 37J, 37K). Characteristic features of Paget's disease include thickened trabeculae, thickened cortex, and sometimes enlarged bone.

Masses arising from within the central sacral canal such as ependymoma, neurogenic tumors, lipoma, and intrasacral meningoceles expand the central spinal canal and erode into the adjacent sacral foramina and body, sometimes causing posterior scalloping (Figs. 37L, 37M). An ependymoma arising from the sacral aspect of the filum terminale may eventually erode through the anterior sacrum into the presacral tissues. Intrasacral meningoceles are un-

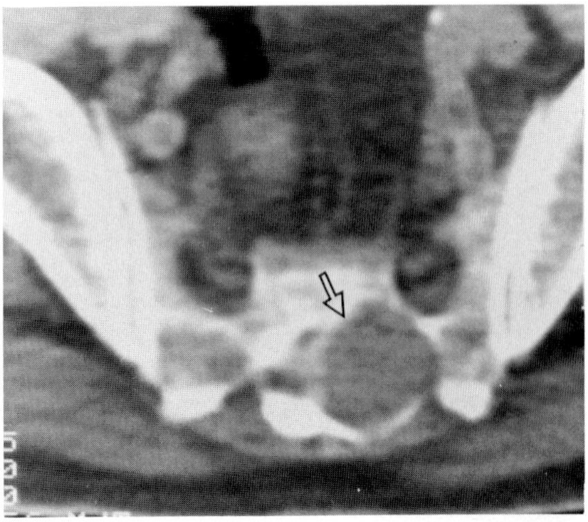

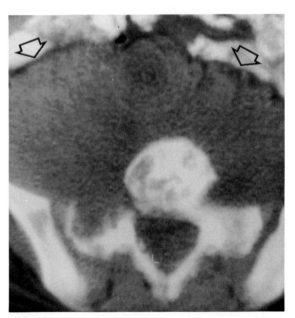

FIG. 37F. Neurilemmoma. There is an eccentrically located osteolytic lesion in the region of the left sacral foramen (*arrow*). A tumor of nerve root or nerve sheath origin may widen the sacral foramen. Although the appearance in this case is not typical, this proved to be a neurilemmoma.

FIG. 37G. Ewing's sarcoma. There is a mixed osteoblastic and osteolytic lesion in the sacrum. Tumor has extended anteriorly forming a huge mass (*arrows*). (Reprinted with permission from Vanel D et al: Computed tomography in the evaluation of 41 cases of Ewing's sarcoma. *Skeletal Radiol* 1982;9:8–13.)

common. CT density measurements demonstrate the fluid nature of the cyst (Fig. 37M). Conventional myelography or CTM can demonstrate communication of the thecal sac with the cyst, although occasionally water-soluble contrast does not enter the cyst.[2]

Lipomas may occur within the sacral canal and may or may not be associated with a subcutaneous component. Spina bifida is almost invariably present. CT can readily distinguish lipoma from other masses because of its negative CT attenuation coefficient.

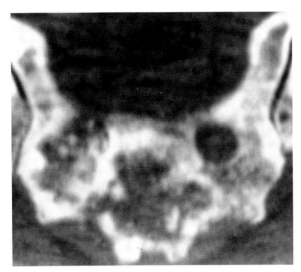

FIG. 37H. Osteosarcoma of sacrum. This patient had had radiation therapy for carcinoma of the cervix 15 years previously. Mottled areas of both osteosclerosis and low attenuation are in part due to radiation osteitis. Tumor has invaded the right sacral foramen.

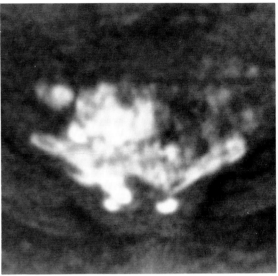

FIG. 37I. Osteosarcoma of the sacrum. This scan was obtained caudal to Figure 37H, at the distal aspect of the sacrum. There is marked bone destruction with ossified tumor extending anteriorly.

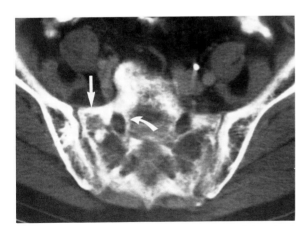

FIG. 37J. Paget's disease. This is an example of the mixed form of Paget's disease with both osteolytic and osteoblastic involvement. Cortical thickening is noted in the anterior sacrum and around the sacral foramen (arrows).

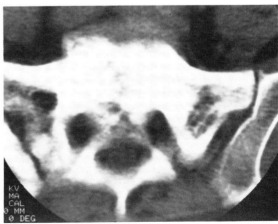

FIG. 37K. Paget's disease. This is the osteoblastic form of this disorder. Trabecular and cortical thickening and diffuse osteosclerosis are evident.

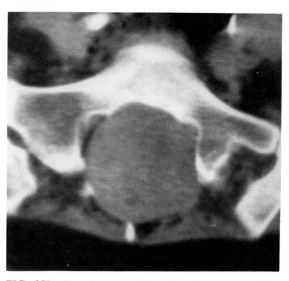

FIG. 37L. Ependymoma. There is a huge mass arising from within the central sacral canal. It has widened the canal and eroded into the sacral foramina and posterior aspect of the sacral body causing scalloping.

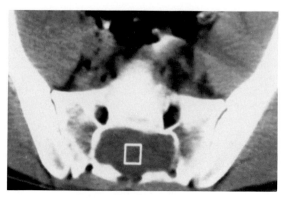

FIG. 37M. Intrasacral meningocele in a 20-year-old male. There is a centrally located osteolytic lesion within the sacrum. The CT attenuation value within the cursor was 7 HU, indicating fluid. This represents an intrasacral meningocele. An aneurysmal bone cyst or unicameral bone cyst could have a similar appearance. (Reprinted with permission from Soye I et al: Computed tomography of sacral and presacral lesions. *Neuroradiology* 1982; 24:71–76.)

References

1. Krol G, Sundaresan N, Deck M: Computed tomography of axial chordomas. *J Comput Assist Tomogr* 1983;7:286–289.
2. Levine E, Batnitzky S: Computed tomography of sacral and presacral lesions. *CRC Crit Rev Diagn Imaging* 1984;21: 307–374.
3. Luken MG III, Michelsen WJ, Whelan MA, et al: The diagnosis of sacral lesions. *Surg Neurol* 1981;15:377–383.
4. Mindell ER, Shah NK, Webster JH: Postradiation sarcoma of bone and soft tissue. *Orthop Clin North Am* 1977; 8:821-834.
5. Rosenthal DI, Scott JA, Mankin HJ, et al: Sacrococcygeal chordoma: Magnetic resonance imaging and computed tomography. *AJR* 1985;145:143–147.
6. Shirkhoda A, Brashear HR, Zelenik ME, et al: Sacral abnormalities—Computed tomography versus conventional radiography. *CT* 1984;8:41–51.
7. Smith J: Radiation-induced sarcoma of bone: Clinical and radiographic findings in 43 patients irradiated for soft tissue neoplasms. *Clin Radiol* 1982; 33:205–221.
8. Smith J, Wixon D, Watson RC: Giant-cell tumor of the sacrum. *J Can Assoc Radiol* 1979;30:34–39.
9. Soye I, Levine E, Batnitzky S, et al: Computed tomography of sacral and presacral lesions. *Neuroradiology* 1982;24:71–76.
10. Vanel D, Contesso G, Couanet D, et al: Computed tomography in the evaluation of 41 cases of Ewing's sarcoma. *Skeletal Radiol* 1982;9:8–13.
11. Weinstein JB, Siegel MJ, Griffith RC: Spinal Ewing sarcoma: Misleading appearances. *Skeletal Radiol* 1984;11:262–265.
12. Whelan MA, Gold RP: Computed tomography of the sacrum: 1. Normal anatomy. *AJR* 1982;139:1183–1190.
13. Whelan MA, Hilal SK, Gold RP, et al: Computed tomography of the sacrum: 2. Pathology. *AJR* 1982;139:1191–1195.
14. Whitehouse GH, Griffiths GJ: Roentgenologic aspects of spinal involvement by primary and metastatic Ewing's tumor. *J Can Assoc Radiol* 1976;27:290–297.

CASE 38

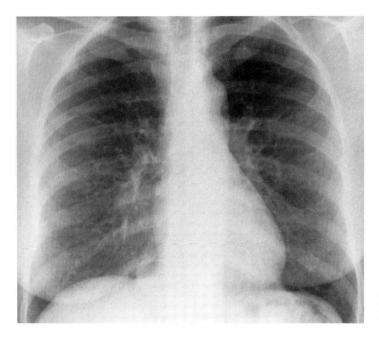

FIG. 38A. PA chest radiograph obtained in a 42-year-old asymptomatic female.

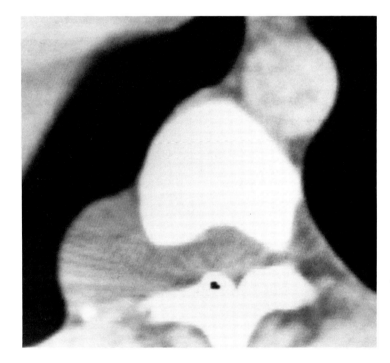

FIG. 38B. Axial CT at the inferior aspect of T10. An abnormality had been identified on the chest x-ray and was further evaluated by CT. This abnormality measured 8 HU. What is the diagnosis?

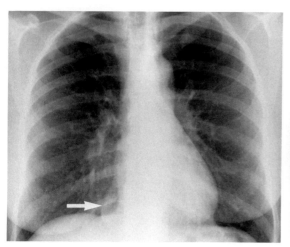 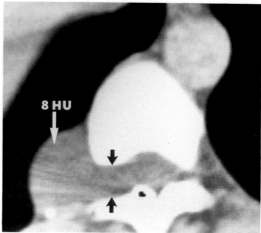

FIG. 38A-1. Lateral thoracic meningocele. There is a paravertebral soft-tissue mass at the T10 and T11 levels on the right (*arrow*).

FIG. 38B-1. The paravertebral mass (*white arrow*) measures 8 HU, representing fluid density. Widening of the neural foramen can be appreciated (*black arrows*). Pantopaque contrast from a myelographic study is noted posteriorly.

Lateral Thoracic Meningocele

In this case a lower thoracic paravertebral mass was detected on routine chest radiography (Fig. 38A-1). A myelogram performed with oil-based contrast demonstrated no intraspinal abnormality but also failed to reveal communication between the subarachnoid space and the mass. On CT examination, a 2.5 × 2.0 cm paravertebral mass is identified with a density measurement of fluid (8 HU) (Fig. 38B-1). Widening of the neural foramen at the T10-T11 level is also noted. The paravertebral mass proved to be a lateral thoracic meningocele.

Lateral thoracic meningocele is formed by protrusion of the dura and arachnoid through an enlarged neural foramen.[6,10] Approximately 60% of lateral thoracic meningoceles are discovered for unrelated reasons (e.g., routine radiographic examination of the chest or radiography of the spine for unrelated symptoms).[9] Patients with lateral thoracic meningocele have associated neurofibromatosis in 65% to 85% of cases.[6,9,11] The etiology of lateral thoracic meningocele is uncertain; however, dural ectasia, faulty attachment of the dura in the intervertebral foramen, elongated nerve root sleeves, bone dysplasia, or a combination of these factors may be the cause.[3,6,9] Lateral thoracic meningoceles are more frequent on the right[9] and usually occur in the midthoracic region, although they may occur at any level.[10] Kyphoscoliosis is found in approximately two thirds of patients with lateral thoracic meningocele, and usually the meningocele is at the apex of curvature on the convex side.[9] Lateral thoracic meningoceles are multiple in 11% of cases and bilateral in 7%.[9] Meningoceles may increase in size over a period of several years.

The differential diagnosis of a paraspinal mass in patients with neurofibromatosis usually includes lateral thoracic meningocele and the less frequent neurofibroma.[4] Unenhanced CT is helpful in establishing a diagnosis of thoracic meningocele and evaluating the extent of the sac. The dilated sac may be detected protruding through a wide neural foramen. Typically the CT density within a lateral thoracic meningocele is similar to that of cerebrospinal fluid, whereas neurofibroma is usually of higher density.[2,5] However, neural tumors such as schwannomas and neurofibromas frequently have relatively low attenuation values (20 to 30 HU).[5,8] In the case of schwannomas this is thought to be due to either the presence of lipid-rich Schwann cells or large amounts of edema, which coalesce to form cystic spaces.[8] Neurofibromas may have low attenuation values because: (1) they may be composed predominantly of lipid-rich Schwann cells, (2) the tumor may contain adipocytes transformed from fibroblasts, or (3) the tumor may undergo central necrosis with cystic degeneration.[8]

CTM may be used to evaluate those problem cases that cannot be accurately diagnosed on conventional CT. Intrathecal water-soluble contrast readily enters a lateral thoracic meningocele; this observation further distinguishes it from a neurofibroma.[1,11] Multiple lateral thoracic meningoceles may occur and can opacify with contrast, thus demonstrating communication with the subarachnoid space[10,11] (Fig. 38C). If need be, further distinction can be made with the use of intravenous contrast. A thoracic meningocele does not enhance after intravenous contrast injection; however, a neurofibroma may demonstrate homogeneous increase of attenuation values.[5] Lateral thoracic meningoceles are also associated with osseous abnormalities such as kyphoscoliosis, enlargement of the neural foramen, erosion of the pedicle, increased interpedicular distance, and vertebral body scalloping.[3,11] Both posterior and lateral vertebral body scal-

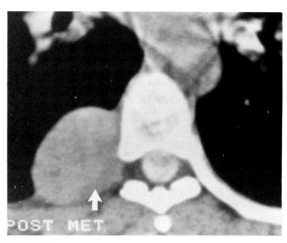

FIG. 38E. Neurofibroma. Same patient as in Fig. 38D. CTM study of the mass followed myelography with water-soluble contrast. There is no communication between the mass (*arrow*) and the well-opacified subarachnoid space. The mass is 4 × 3.5 cm in size and has a CT density measurement of 46 HU.

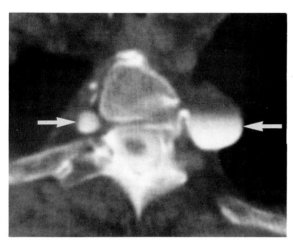

FIG. 38C. Multiple lateral thoracic meningoceles. Axial CTM demonstrates bilateral contrast-filled lateral thoracic meningoceles (*arrows*). (Reprinted with permission from Weinreb JC et al: CT metrizamide myelography in multiple bilateral intrathoracic meningoceles. *J Comput Assist Tomogr* 1984;8:324–326.)

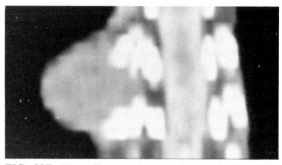

FIG. 38F. Neurofibroma. Same patient as in Fig. 38D. Coronal reconstruction of axial CTM images through the plane of the pedicles and mass clearly demonstrates lack of communication between the subarachnoid space and the neurofibroma.

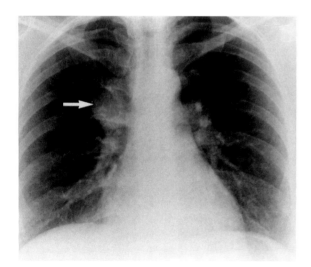

FIG. 38D. Neurofibroma of an intercostal nerve. A PA chest radiograph demonstrates a mass in the posterior mediastinum on the right (*arrow*). This 37-year-old female had had chest discomfort for 5 months. She had no history of neurofibromatosis.

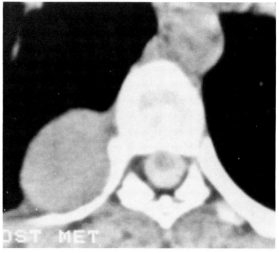

FIG. 38G. Neurofibroma. Same patient as in Fig. 38D. Axial CTM. The mass is causing rib erosion, a typical feature of neurofibroma of an intercostal nerve.

loping have been described in association with lateral thoracic meningocele. The bony erosions may include the posterior elements and if severe enough may require posterior fusion for stability.[7] The extent of osseous erosion is best identified by CT.

A patient with a neural tumor arising from an intercostal nerve or nerve sheath may present with a paravertebral mass identified on chest radiography (Fig. 38D). CTM demonstrates no communication with the subarachnoid space (Figs. 38E, 38F), although rib erosion or other osseous abnormalities may be seen (Fig. 38G).

References

1. Angtuaco EJC, Binet EF, Flanigan S: Value of computed tomographic myelography in neurofibromatosis. *Neurosurgery* 1983;13:666–671.
2. Biondetti PR, Vigo M, Fiore D, et al: CT appearance of generalized von Recklinghausen neurofibromatosis. *J Comput Assist Tomogr* 1983;7:866–869.
3. Booth AE: Lateral thoracic meningocele. *J Neurol Neurosurg Psychiatry* 1969;32:111–115.
4. Casselman ES, Miller WT, Lin SR, et al: Von Recklinghausen's disease: Incidence of roentgenographic findings with a clinical review of the literature. *CRC Crit Rev Diagn Imaging* 1977;9:387–419.
5. Coleman BG, Arger PH, Dalinka MK, et al: CT of sarcomatous degeneration in neurofibromatosis. *AJR* 1983;140:383–387.
6. Erkulvrawatr S, Gammal TE, Hawkins J, et al: Intrathoracic meningoceles and neurofibromatosis. *Arch Neurol* 1979;36:557–559.
7. Kornberg M, Rechtine GR, Depuy TE: Thoracic vertebral erosion secondary to an intrathoracic meningocele in a patient with neurofibromatosis. *Spine* 1984;9:821–824.
8. Kumar AJ, Kuhajda FP, Martinez CR, et al: Computed tomography of extracranial nerve sheath tumors with pathological correlation. *J Comput Assist Tomogr* 1983;7:857–865.
9. Miles J, Pennybacker J, Sheldon P: Intrathoracic meningocele: Its development and association with neurofibromatosis. *J Neurol Neurosurg Psychiatry* 1969;32:99–110.
10. O'Neill P, Whatmore WJ, Booth AE: Spinal meningoceles in association with neurofibromatosis. *Neurosurgery* 1983;13:82–84.
11. Weinreb JC, Arger PH, Grossman R, et al: CT metrizamide myelography in multiple bilateral intrathoracic meningoceles. *J Comput Assist Tomogr* 1984;8:324–326.

CASE 39

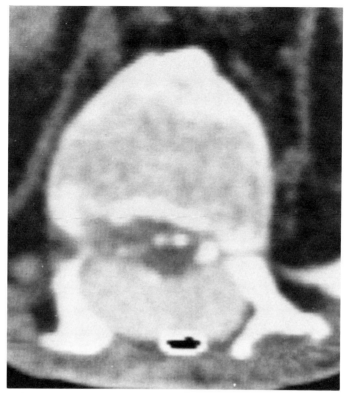

FIG. 39A. Axial CT of the thoracolumbar spine after intrathecal introduction of water-soluble contrast. This 36-year-old man has had a systemic disorder since age 8 and now has severe back pain.

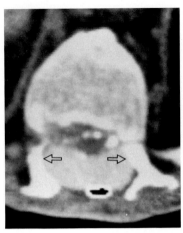

FIG. 39A-1. Neurofibromatosis. The pedicles are hypoplastic. Dural ectasia is present causing mesial scalloping of the pedicles (*arrows*). In addition there is spina bifida. These abnormalities are due to neurofibromatosis. A droplet of Pantopaque contrast is seen posteriorly and is from a previous myelogram.

Neurofibromatosis

The axial CT scan of the lower thoracic spine demonstrates posterior scalloping of the vertebral body, dural ectasia, hypoplasia and mesial scalloping of the pedicles, and spina bifida (Fig. 39A-1). This combination of findings is due to neurofibromatosis. There is also compression of the subarachnoid space (Figs. 39B, 39C) due to severe kyphoscoliosis of the thoracolumbar spine (Figs. 39B, 39D). A paravertebral mass is present and is associated with lateral scalloping of the vertebral body (Figs. 39C, 39D).

Neurofibromatosis is a hereditary hamartomatous disorder, probably of neural crest origin, involving neuroectodermal, mesodermal, and endodermal tissues.[4] It is classified as one of the phacomatoses.[3] Spinal alterations are frequent in neurofibromatosis and are most often due to mesodermal dysplasia (bone, dura) but may also be caused by neural tumors and lateral meningoceles.[4,5] In the general population, the most common cause of posterior scalloping of one or more vertebral bodies is neurofibromatosis.[5] Posterior scalloping is present in approximately 14% of patients with neurofibromatosis and is most often due to dural ectasia.[3] The dural ectasia occurs secondary to mesenchymal abnormality, which allows dilatation of the subarachnoid space.[5] Posterior scalloping can also occur secondary to one or more neurofibromas. Although posterior scalloping of the vertebral body can be demonstrated by conventional radiography, CTM and myelography are useful modalities capable of determining the cause of the posterior vertebral scalloping (i.e., distinguishing between dural ectasia and tumor).[1] Lateral and anterior vertebral body scalloping may occur either secondary to adjacent neural tumor or as a primary mesodermal dysplasia of bone.[2]

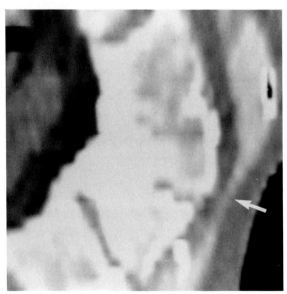

FIG. 39B. Neurofibromatosis. Same patient as in Fig. 39A-1. Sagittal reconstruction of axial CTM images. Kyphosis, most severe from T11 to L2, is causing marked compression of the subarachnoid space (*arrow*).

Kyphoscoliosis is the most frequent skeletal abnormality in neurofibromatosis and occurs in approximately 45% of patients.[3] Again, this is usually the result of mesodermal dysplasia but in some cases may be related to thoracolumbar spinal tumors.[3] When severe enough, as in this case, kyphoscoliosis may cause spinal cord or cauda equina compression. Symptoms of cord compression in a patient with neurofibromatosis can be evaluated by myelography and CTM. By these methods, symptoms referable to kyphoscoliosis

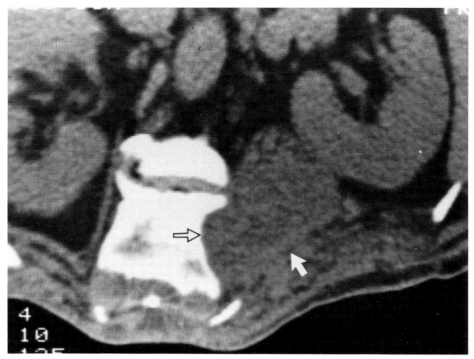

FIG. 39C. Neurofibromatosis. Axial CTM in the same patient as in Fig. 39A-1. There is marked compression of the thecal sac at this level due to kyphosis. No mass is present within the spinal canal; however, a low-density soft-tissue mass is seen in the paravertebral region (*closed arrow*). This paravertebral mass is causing lateral scalloping of the vertebral body (*open arrow*). The nature of this mass is not proven but it is most likely a neural tumor.

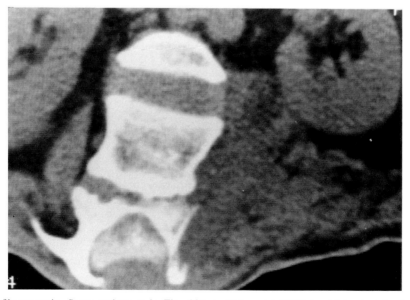

FIG. 39D. Neurofibromatosis. Same patient as in Fig. 39A-1. This axial CTM is obtained at the level of maximum kyphosis. Because of the severe kyphosis, three vertebrae and two intervertebral disc spaces are examined on the same axial CT section. Posterior scalloping of the vertebral body and mesial scalloping of the pedicles can be appreciated. The left paravertebral mass is again noted.

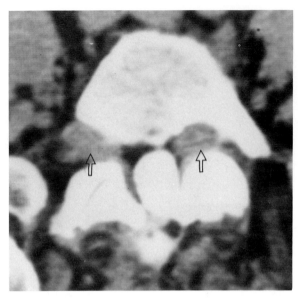

FIG. 39E. Neurofibromatosis. Bilateral neurofibromas (*arrows*) are present at L5–S1 in this patient with neurofibromatosis. Posterior vertebral scalloping is noted on the left.

alone can be distinguished from those due to tumor or a combination of tumor and spinal deformity. One or more pedicles may be abnormal in patients with neurofibromatosis. Agenesis, hypoplasia, and mesial erosion of the pedicles have been described[3,6] (Figs. 39A-1, 39D). Other skeletal manifestations of neurofibromatosis include widening of the neural foramen, widening of the spinal canal, hypoplasia of the transverse process, spina bifida, and thinning of the ribs.[3,4,7]

The presence of a paraspinal mass in association with neurofibromatosis has been described in a previous case and is usually due to a lateral meningocele or less often a neural tumor. CTM is used to distinguish a lateral meningocele from a neural tumor and thus helps to direct therapy.[1] In the present case, there is no surgical confirmation of the paravertebral mass; however, it most likely represents a neural tumor. Another patient with neurofibromatosis and bilateral neurofibromas is shown in Fig. 39E.

References

1. Angtuaco EJC, Binet EF, Flanigan S: Value of computed tomographic myelography in neurofibromatosis. *Neurosurgery* 1983;13:666–671.
2. Casselman ES, Mandell GA: Vertebral scalloping in neurofibromatosis. *Radiology* 1979;131:89–94.
3. Casselman ES, Miller WT, Lin SR, et al: Von Recklinghausen's disease: Incidence of roentgenographic findings with a clinical review of the literature. *CRC Crit Rev Diagn Imaging* 1977;9:387–419.
4. Holt JF: Neurofibromatosis in children. *AJR* 1978;130:615–639.
5. Leeds NE, Jacobson HG: Spinal neurofibromatosis. *AJR* 1976;126:617–623.
6. Mandell GA: The pedicle in neurofibromatosis. *AJR* 1978;130:675–678.
7. Yaghmai I: Spine changes in neurofibromatosis. *Radiographics* 1986;6:261-285.

CASE 40

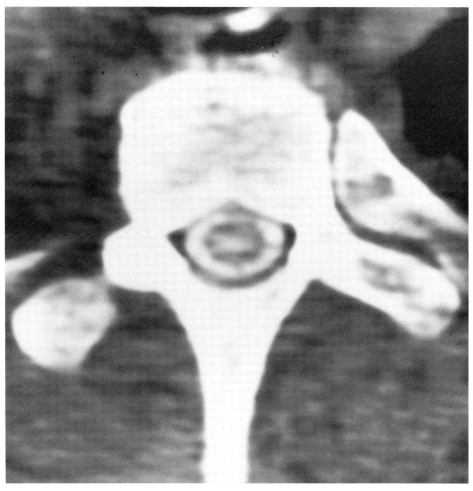

FIG. 40A. Axial CT scan at T2 performed several hours after introduction of intrathecal water-soluble contrast. This patient had had previous cervical spine surgery and now presented with progressive myelopathy.

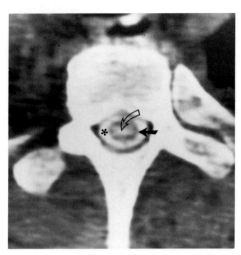

FIG. 40A-1. Syringomyelia. Water-soluble contrast fills a syrinx (*curved arrow*) within a normal-sized spinal cord (*straight arrow*). Contrast can also be seen within the surrounding subarachnoid space (asterisk). Additional scans demonstrated the extent of the syrinx, which involved portions of the cervical and thoracic cord.

Syringomyelia

A CT study performed several hours after the intrathecal introduction of water-soluble contrast demonstrates contrast within the spinal cord (Fig. 40A-1). This is due to syringomyelia, an intramedullary cavity lined by glial cells that develops extrinsic to the central canal of the spinal cord. Another form of intramedullary cavity is hydromyelia, which is lined by ependymal cells and represents cystic dilatation of the central canal of the spinal cord. Hydromyelia is of congenital origin and is frequently associated with hindbrain abnormalities such as Chiari malformation.[3] Some patients with hydromyelia may develop syringomyelia. Because syringomyelia and hydromyelia are so closely associated developmentally, clinically, and even radiographically, they may be considered together as syringohydromyelia.[2] Usually, syringohydromyelia involves the dorsal aspect of the cervical cord. The cavity in some cases may extend into the thoracic region and even involve the entire cord. Patients with syringohydromyelia typically present with diminution of pain and temperature perception in the upper extremities, which progresses to amyotrophy and paresis.[2] Thoracic scoliosis and neurotrophic joints in the upper extremities also occur in this disorder.

In the evaluation of patients with syringohydromyelia, conventional radiography may demonstrate widening of the spinal canal. Conventional myelography may reveal an enlarged cord with narrowing of the subarachnoid space, similar in appearance to an intramedullary tumor. Myelography performed with water-soluble iodinated contrast or gas can establish a diagnosis of syringohydromyelia by demonstrating collapse of the spinal cord with a change in patient position.[5] Opacification of the cyst is not detected by myelography.

The CT appearance of syringohydromyelia varies. Studies of cord size in patients with syringohydromyelia have had differing results; however, in one large series of adults examined by CTM, spinal cord size was most frequently normal (45%) or reduced (45%); it was enlarged in only 10% of cases[1] (Fig. 40B). Others have noted that cord size is usually enlarged in children.[7] The shape of the cord may be round, oval, or flat (Fig. 40C). A flattened cord that changes shape with change in patient position is diagnostic of syringohydromyelia[10] but is more easily demonstrated with myelography. The cyst itself is sometimes visualized by unenhanced CT by virtue of its low attenuation value, which is similar to that of cerebrospinal fluid and on average approximately 15 HU less than that of the spinal cord.[8] Cysts that are isodense with the cord or are small and collapsed may go undetected by unenhanced CT.[1] CTM, however, is an excellent modality for diagnosis and evaluation of syringohydromyelia. Although only a small percentage of cysts are opacified on the scan performed immediately following the intrathecal injection of contrast, most cysts are opacified on the delayed scan obtained 6 to 10 hours later[1] (Figs. 40D, 40E). In addition, CTM can detect Chiari malformation associated with hydromyelia.[1] Hydrocephalus can be diagnosed during the CT examination.

Syringomyelia is sometimes "acquired," developing secondary to spinal cord tumor,[2-4] trauma and hematomyelia,[9,11,12] or arachnoiditis.[13] When a

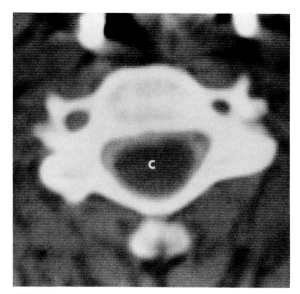

FIG. 40B. Syringomyelia. CTM at C5 performed immediately following intrathecal injection of water-soluble contrast. There is uniform enlargement of the spinal cord (*C*). The subarachnoid space is narrowed by the expanded cord. The intramedullary cavity did not fill on this immediate CTM study but did opacify on the delayed CTM examination. This 54-year-old presented with numbness and sensory loss in the right upper extremity. (Figure courtesy of Norman Komer, M.D., Tucson, Ariz.)

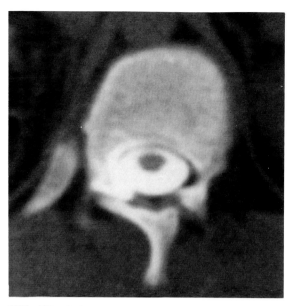

FIG. 40D. Syringohydromyelia. This 14-year-old male had a myelomeningocele repaired shortly after birth. CTM at a lower thoracic vertebral level immediately following injection of intrathecal contrast. No contrast is noted within the small spinal cord.

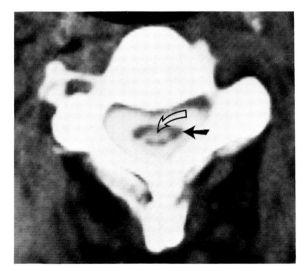

FIG. 40C. Syringomyelia. CTM at the C3 level. The spinal cord (*straight arrow*) has lost its normal round or elliptical shape and has a somewhat flat appearance. Water-soluble contrast is present within the syrinx cavity (*curved arrow*). This 70-year-old woman had weakness of both arms and loss of pain sensation. (Figure courtesy of Richard Wachter, M.D., Tucson, Ariz.)

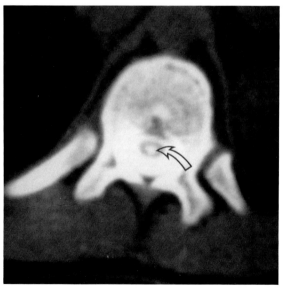

FIG. 40E. Syringohydromyelia. Delayed CTM scan of the same patient as in Fig. 40D. Note that water-soluble contrast has entered the intramedullary cavity (*arrow*) on this delayed CTM study. (Figs. 40D and 40E courtesy of Spencer Borden IV M.D., Philadelphia, Pa.).

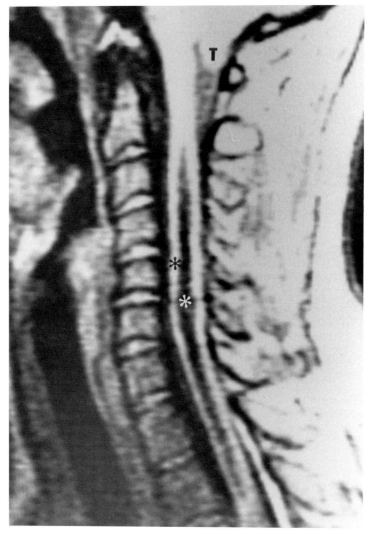

FIG. 40F. Syringohydromyelia. T1-weighted MRI of the cervical spine in the sagittal projection. The fluid in the cavity (*white asterisk*) is contrasted with the brighter signal of the spinal cord (*black asterisk*). Note the extensive nature of the intramedullary cavity, which is readily evaluated with MRI. Also note the inferior displacement of the cerebellar tonsils (*T*) in this patient with Chiari I malformation.

syrinx cavity cannot be demonstrated at all levels of cord enlargement, an associated tumor may be suspected.[4] Tumor should also be considered when the syrinx spares the cervical cord (the typical location of "true" syringomyelia), extends for only a short segment, or has a nodular, irregular wall.[8] Compared to the spinal cord, a cyst associated with spinal cord tumor is usually isodense because of an elevated protein content or hypodense. Occasionally the attenuation value of the cyst is elevated (e.g., in cases of dermoid cyst).[8] A low CT density within a tumor need not indicate a cyst since edema and gliosis may exist within the tumor, rendering it hypodense. Although some authors state that intrathecal water-soluble contrast uncommonly opacifies a syrinx associated with tumor,[8] others have demonstrated opacification of the cavity with delayed scanning obtained 7 to 24 hours after initial installation of contrast.[4] In one case described, opacification was demonstrated on the 24-hour scan although no opacification had been seen 8 hours after contrast introduction.[4] Intravenous contrast may be used to enhance a tumor; however, the absence of enhancement certainly does not exclude tumor.[4,8]

Syringomyelia may develop following significant spinal cord trauma. It usually presents with new and/or progressively deteriorating neurologic symptoms, which begin months to many years after the injury.[9] Less frequently, there is progressive deterioration in the neurologic status from the onset of injury, and in some cases there is a failure of new or progressive symptoms to develop. A cyst is formed by liquefaction of the hemorrhagic necrosis at the site of injury and from extension of necrosis into adjacent normal cord.[14] These cysts usually opacify on the delayed CT scan, but occasionally they may not.[11,12] Opacification of cysts is seen more frequently in enlarged cords.[14] Cysts may be small or may involve the entire cord. Multiple cysts can occur.[9] Syringomyelia may develop secondary to arachnoiditis. CTM can readily demonstrate the syrinx and can also detect findings of adhesive arachnoiditis.[13]

Magnetic resonance imaging (MRI) is useful in demonstrating syringomyelia in the sagittal plane[6] (Fig. 40F). Associated Chiari malformation is readily identified by MRI. In addition, MRI can be used to distinguish tumor from the syrinx by differences in signal characteristics.[6]

References

1. Aubin ML, Vignaud J, Jardin C, et al: Computed tomography in 75 clinical cases of syringomyelia. *AJNR* 1981;2:199–204.
2. Ballantine HT Jr, Ojemann RG, Drew JH: Syringohydromyelia. *Progr Neurol Surg* 1971;4:227–245.
3. Gardner WJ: Hydrodynamic mechanism of syringomyelia: Its relationship to myelocele. *J Neurol Neurosurg Psychiatry* 1965;28:247–259.
4. Kan S, Fox AJ, Viñuela F, Barnett HJM, et al: Delayed CT metrizamide enhancement of syringomyelia secondary to tumor. *AJNR* 1983;4:73–78.
5. Kan S, Fox AJ, Viñuela F, et al: Spinal cord size in syringomyelia: Change with position on metrizamide myelography. *Radiology* 1983;146:409–414.
6. Lee BCP, Zimmerman RD, Manning JJ, et al: MR imaging of syringomyelia and hydromyelia. *AJNR* 1985;6:221-228.
7. Pettersson H, Harwood-Nash DCF: *CT and Myelography of the Spine and Cord: Techniques, Anatomy and Pathology in Children*. Berlin, Springer-Verlag, 1982.
8. Pullicino P, Kendall BE: Computed tomography of "cystic" intramedullary lesions. *Neuroradiology* 1982;23:117–121.
9. Quencer RM, Green BA, Eismont FJ: Posttraumatic spinal cord cysts: Clinical features and characterization with metrizamide computed tomography. *Radiology* 1983;146:415–423.
10. Resjö IM, Harwood-Nash DC, Fitz CR, et al: Computed tomographic metrizamide myelography in syringohydromyelia. *Radiology* 1979;131:405–407.
11. Rossier AB, Foo D, Naheedy MH, et al: Radiography of posttraumatic syringomyelia. *AJNR* 1983;4:637–640.
12. Seibert CE, Dreisbach JN, Swanson WB, et al: Progressive posttraumatic cystic myelopathy: Neuroradiologic evaluation. *AJNR* 1981;2:115–119.
13. Simmons JD, Norman D, Newton TH: Preoperative demonstration of post inflammatory syringomyelia. *AJNR* 1983;4:625–628.
14. Stevens JM, Olney JS, Kendall BE: Post-traumatic cystic and non-cystic myelopathy. *Neuroradiology* 1985;27:48–56.

CASE 41

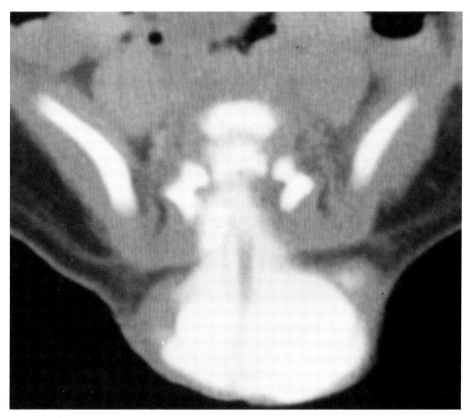

FIG. 41A. CTM scan of a 6-day-old infant with a large, fluctuant, palpable mass in the lumbosacral region.

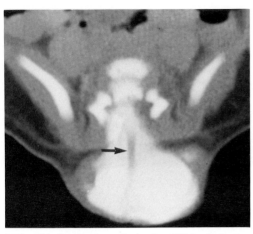

FIG. 41A-1. Myelomeningocele. CTM of the lower lumbar spine in the axial plane. There is protrusion of the contrast-filled thecal sac through a wide spina bifida. Note the angle of the laminae, which appear flared. Neural tissue (*arrow*) is identified within the sac, thus classifying this lesion as a myelomeningocele.(Figure courtesy of Derek Armstrong, M.D., Toronto, Canada.)

Myelomeningocele

The contrast-filled subarachnoid space extends posteriorly into the subcutaneous tissue through a wide defect in the laminae (spina bifida) (Fig. 41A-1). Neural tissue is present within the large posterior sac. This is a myelomeningocele—a form of spinal dysraphism.

Spinal dysraphism indicates a failure of complete fusion of tissues in the dorsal median plane of the developing embryo that leads to anomalies of the skin, bones, dura, spinal cord, and nerves. Spinal dysraphism is a complex clinical state associated with a broad spectrum of abnormalities that vary in severity and may occur alone or in combination. Some dysraphic conditions such as myeloceles, myelomeningoceles, and many meningoceles are overt—that is, clearly visible and easily diagnosed on physical examination. Other forms of dysraphism, such as lipomyelomeningocele, diastematomyelia, tethered cord, and associated intraspinal masses such as lipomas and cysts, are occult—that is, not visible on physical examination but suspected because of cutaneous and/or neurologic, orthopedic, or urologic abnormalities.[10] The cutaneous anomalies discovered in patients with occult dysraphism usually develop in the lower back and include a patch of hair, nevus, lipoma, or dermal sinus.[6] Occasionally, occult spinal dysraphism may go undiagnosed until adult life.[11] In the overall classification of these disorders, we prefer, as do others, the term spinal dysraphism to spina bifida. Spina bifida (bifid spine) is considered a skeletal dysraphism in which there is a fusion defect of the posterior vertebral elements or, rarely, the vertebral body (Fig. 41B). The defect varies from a narrow slit of the lamina, detected only by radiography, to splaying or absence of the laminae at several levels. Mild, narrow posterior spina bifida, particularly at L5-S1, is commonly found in asymptomatic patients and is called spina bifida occulta. It occurs in approximately 20% of the general population and by itself is of no clinical significance.[1]

Meningocele is a form of dysraphism in which there is herniation of skin-covered arachnoid and dura through a spina bifida, most often in the lumbosacral region.[3] Neural tissue is not present within the protruding sac.[6] Vertebral arch defects and widening of the spinal canal are localized and relatively mild. There is no association with syringohydromyelia or Chiari malformation.[6]

A myelocele is a severe form of dysraphism in which a plaque of neural tissue representing malformed spinal cord lies exposed and flush with the skin surface.[6] Ventral and dorsal nerve roots arising from the neural plaque traverse the subarachnoid space. A myelomeningocele is a myelocele that has been elevated above the skin surface by a protruding arachnoid space (meningocele)[3,6] (Fig. 41A-1). The term myelomeningocele is often used to include both myeloceles and myelomeningoceles (i.e., myelomeningocele without or with a sac) and is similarly used here. Myelomeningoceles usually develop in the lumbosacral region. These disorders are more devastating than simple meningoceles. Patients with myelomeningocele almost invariably have tethered cord and Chiari II malformation.[6] They frequently have hydrocephalus and less often have associated diastematomyelia and/or syringohydromyelia.[6] Dysraphic changes in the

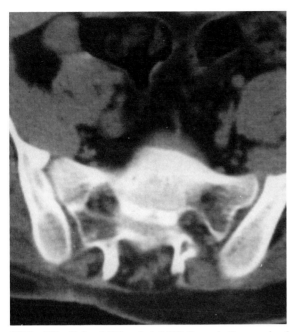
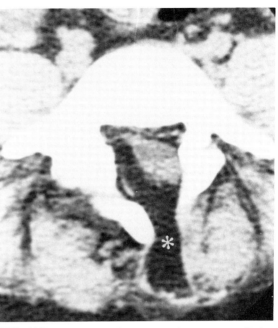

FIG. 41B. Spina bifida. Axial CT through the proximal sacrum. The laminae are widely splayed, forming a large dorsal bony defect.

FIG. 41C. Lipoma. There is a large lipoma (*asterisk*) that is both intraspinal and extraspinal, traversing through a spina bifida. This patient has an associated tethered cord, not shown in this section.

spine are more extensive than those that occur with simple meningocele.

Lipomyelomeningocele is a skin-covered myelomeningocele associated with a subcutaneous lipomatous-connective tissue mass (lipoma). The lipoma attaches to the dorsal surface of the neural plaque and tethers the cord.[7] Skeletal manifestations of dysraphism are evident. The lipoma may extend into the spinal canal through the dorsal defect.[7] Lipomyelomeningoceles are likely to be associated with a neuromusculoskeletal syndrome at birth.[3] Chiari malformation, hydrocephalus, syringohydromyelia, and diastematomyelia are not usually associated with lipomyelomeningocele.[6]

Similar subcutaneous lipomatous masses associated with dorsal dysraphism may exist without a meningocele[7] and are the most frequent form of occult spinal dysraphism[3] (Fig. 41C). They extend through a spina bifida of variable size and also attach to and tether the cord.[3] These lipomas are not associated with other forms of spinal dysraphism. They usually present in infancy or childhood.[3] There are other intraspinal lipomas that remain almost completely within a normal or mildly bifid spinal canal.[6] Although these intraspinal lipomas consist of normal fat cells[7] and are not true neoplasms, they may compress neural tissue within the spinal canal by their mass effect.

In patients with occult or overt spinal dysraphism, the decision as to which imaging modalities are utilized varies with the disorder, the clinical condition of the patient, the institution, the imaging modalities available, and the experience of the imaging physician. Whenever occult dysraphism is suspected radiographic studies should be obtained and the patient treated as soon as possible.[10] On the other hand, most meningoceles are easily diagnosed at birth and, in healthy infants, may be repaired during the first week of life.[3] Since they do not as a rule have a tethered cord or Chiari malformation, and surgery is relatively uncomplicated, imaging other than conventional radiography is not usually performed even though the thecal sac is readily accessible.[2,3] However, on occasion a fluctuant mass in the low back may not be a fluid-filled meningocele but rather a lipoma, cystic tumor, or other cystic lesion[3] (Fig. 41D). When the diagnosis of meningocele is uncertain, imaging with ultrasonography, myelography, and/or CTM can aid in establishing a diagnosis.

Myelomeningocele is easily diagnosed clinically at birth. Most patients are not imaged preoperatively since surgery is performed on viable children as soon as possible (within 30 to 36 hours after birth),[3] and it is known that patients with myelomeningocele almost invariably have a tethered cord and Chiari II malformation. There is also difficulty in injecting

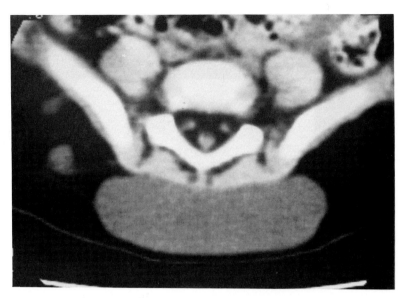

FIG. 41D. Lymphangioma. This 6-month-old child presented with a palpable mass in the low back that had gradually increased in size since birth. CTM at a lower lumbar level reveals a large mass of fluid density in the subcutaneous tissues. This mass did not fill with intrathecal contrast. The laminae are intact, precluding a dysraphic condition. (Figure courtesy of Spencer Borden IV, M.D., Philadelphia, Pa.)

contrast intrathecally in these patients. The goal of surgery is to free the neural plaque, place it into the spinal canal, and reconstruct the dura.[2,10]

Whenever imaging is desired, conventional radiographs of the entire spine in the AP and lateral projections provide a survey for many skeletal dysraphic alterations. Spina bifida, wide sagittal diameter of the spinal canal, wide interpedicular distance, abnormal vertebral segmentation (butterfly vertebra and hemivertebra), bony spur (in diastematomyelia), fused laminae, and other anomalies give clues to the possible location of underlying associated dysraphic abnormalities. This is important because dysraphic disorders may exist at multiple levels. In newborns with incomplete ossification of posterior elements, mild dysraphic alterations in the spine may be difficult to detect with conventional radiography. Conventional myelography can readily outline a protruding sac and the level of the spinal cord. It is an adequate diagnostic modality for evaluation of the entire subarachnoid space and associated intraspinal abnormalities. This guides the imaging physician to the areas of interest.

CTM is an excellent modality for evaluation of dysraphic disorders and offers more information than either conventional myelography or unenhanced CT.[9] It can accurately delineate the caudal extent of the spinal cord and the presence and location of the neural plaque and nerve root in the protruding sac (Fig. 41A-1). With the use of serial sections, the dorsally positioned spinal cord can be followed as it enters the sac and attaches to the sac wall as the neural plaque.[2] CTM can demonstrate the presence of associated disorders such as diastematomyelia, syringohydromyelia, Chiari malformation, and intraspinal or extraspinal masses or cysts. Fatty masses can be diagnosed by virtue of their negative CT attenuation values. The nature and extent of skeletal deformities can be determined, and the brain can also be easily examined for hydrocephalus. CT is helpful in the follow-up evaluation of patients who develop complications following surgical correction of myelomeningocele. Some postsurgical patients develop scoliosis and progressive spasticity of the extremities. Hydromyelia and/or compression of the brainstem by Chiari malformation may be responsible for the delayed neurological deterioration and can be detected with CTM.[8] Tethered cord may develop following myelomeningocele repair and can also be evaluated by CTM.[2]

Ultrasonography can be utilized in selected patients as a screening modality for (lipo-, myelo-) meningocele.[5] It can delineate the size and shape of the protruding sac and its relationship to the spinal column and skin surface. Tethered cord, lipoma, and to a lesser extent nerve roots can also be detected. Magnetic resonance imaging (MRI) is an excellent imaging modality for the evaluation of some dysraphic conditions. The entire spinal cord, brainstem, fourth ventricle, and cerebellum can be readily demonstrated so that syringohydromyelia, tethered cord, Chiari malformation, and meningocele can be evaluated.[4] MRI can differentiate lipoma from other intraspinal masses.

References

1. Boone D, Parsons D, Lachmann SM, et al: Spina bifida occulta: Lesion or anomaly? *Clin Radiol* 1985;36:159–161.
2. Fitz CR: Midline anomalies of the brain and spine. *Radiol Clin North Am* 1982;20:95–104.
3. French BN: Midline fusion defects and defects of formation, in Youmans JR (ed): *Neurological Surgery*, vol 3. Philadelphia, WB Saunders, 1982, pp 1236–1380.
4. Han JS, Kaufman B, El Yousef SJ, et al: NMR imaging of the spine. *AJNR* 1983;4:1151–1159, *AJR* 1983;141:1137–1145.
5. Naidich TP, Fernbach SK, McLone DG, et al: Sonography of the caudal spine and back: Congenital anomalies in children. *AJR* 1984;142:1229–1242.
6. Naidich TP, McLone DG, Harwood-Nash DC: Spinal dysraphism, in Newton TH, Potts DG (ed): *Computed Tomography of the Spine and Spinal Cord*. San Anselmo, Calif, Clavadel Press, 1983, pp 299–353.
7. Naidich TP, McLone DG, Mutluer S: A new understanding of dorsal dysraphism with lipoma (lipomyeloschisis): Radiologic evaluation and surgical correction. *AJR* 1983;140:1065–1078.
8. Parks TS, Cail WS, Maggio WM, et al: Progressive spasticity and scoliosis in children with myelomeningocele. *J Neurosurg* 1985;62:367–375.
9. Resjö IM, Harwood-Nash DC, Fitz CR, et al: Computed tomographic metrizamide myelography in spinal dysraphism in infants and children. *J Comput Assist Tomogr* 1978;2:549–558.
10. Schut L, Bruce DC, Sutton LN: The management of the child with lipomyelomeningocele. *Clin Neurosurg* 1983;30:464–476.
11. Sostrin RD, Thompson JR, Rouhe SA, et al: Occult spinal dysraphism in the geriatric patient. *Radiology* 1977;125:165–169.

CASE 42

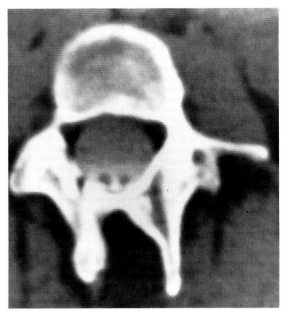

FIG. 42A. This 38-year-old woman presented with an 8-month history of intermittent pain in the right leg and foot as well as disturbance of gait. Axial CT scan at the L2 level following introduction of intrathecal water-soluble contrast.

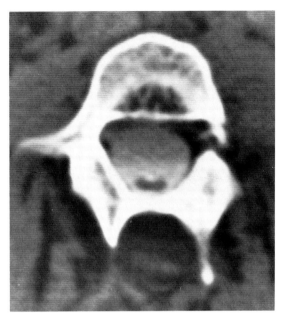

FIG. 42B. Axial CTM at L3.

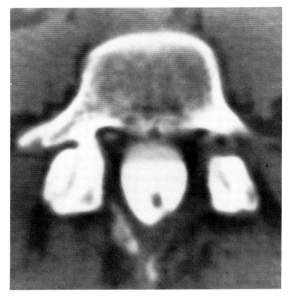

FIG. 42C. Axial CTM at L5.

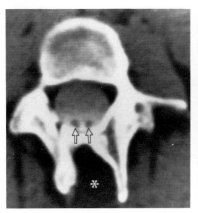 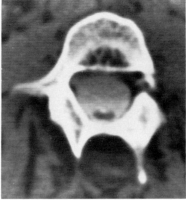 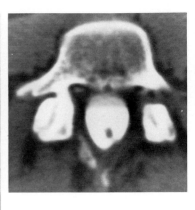

FIG. 42A-1. Diastematomyelia. There are two round filling defects (*arrows*) representing two hemicords that are dorsally situated within the contrast-filled thecal sac. This represents diastematomyelia. Spina bifida is present (asterisk).

FIG. 42B-1. The two hemicords are merging at this level.

FIG. 42C-1. A single filling defect is seen within the thecal sac, which at this lower lumbar level represents either a tethered cord or a thickened filum terminale.

Diastematomyelia

CTM demonstrates the spinal cord split into two small hemicords in the upper lumbar spine (Figs. 42A-1, 42B-1). Splitting of the spinal cord is termed diastematomyelia. An associated spina bifida is present. No bony septum is demonstrated. Tethering of the cord or thickening of the filum terminale is present more caudad at the lower lumbar levels (Fig. 42C-1). Coronal reconstruction through the plane of the two hemicords again demonstrates the abnormality (Fig. 42D).

Diastematomyelia is a form of occult spinal dysraphism in which the spinal cord, conus medullaris, and/or filum terminale are partially or completely divided sagittally into two nearly equal hemicords.[6] Each hemicord contains a central canal and an ipsilateral dorsal and ventral horn. The two hemicords usually rejoin caudally into a single cord.[4] In 50% to 80% of cases of diastematomyelia, the two hemicords lie within a common subarachnoid space enclosed within a single arachnoid and dura.[6,8,9] This form of diastematomyelia has no fibrous septum or osseous cartilaginous spur separating the two hemicords[6] (Fig. 42A-1). In the other 20% to 50% of cases, the two hemicords each lie within their own separate subarachnoid space covered by their own separate arachnoid and dura.[6,8,9] This form of diastematomyelia nearly always has a fibrous septum or osseous cartilaginous spur separating the two dural tubes, each of which contains a hemicord[6] (Figs. 42E, 42F). The fibrous septum or bony spur may tether the spinal cord.

Clinically, the signs and symptoms of diastematomyelia may be similar to those of other forms of spinal dysraphism and frequently include a cutaneous abnormality of the back such as a patch of hair, skin dimple, or nevus; abnormal gait; leg weakness; abnormal lower extremity reflexes; scoliosis; and urinary and fecal incontinence.[4,9] In one large series,

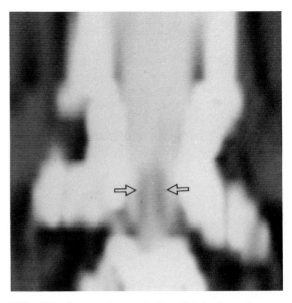

FIG. 42D. Same patient as in Fig. 42A-1. Coronal reconstruction of axial images through the posterior spinal canal. The two hemicords are visualized in part (*arrows*). The lumbar lordosis prevents imaging the entire extent of the diastematomyelia on a single coronal reconstruction view.

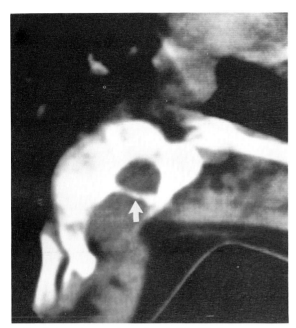

FIG. 42E. Diastematomyelia. Axial CT demonstrates a bony septum (*arrow*) dividing the spinal canal. Spina bifida and severe scoliosis are present.

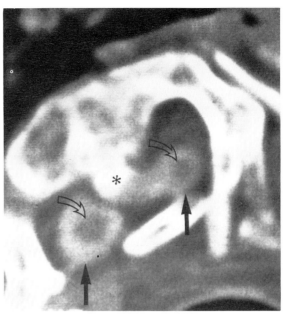

FIG. 42F. Diastematomyelia. Same patient as in Fig. 42E. CTM demonstrates two separate thecal sacs (*straight arrows*) within which lie two separate hemicords (*curved arrows*). A portion of the bony septum is noted (asterisk). (Figs. 42E and 42F courtesy of Gerald Mandell, M.D., Wilmington, Del.)

diastematomyelia was found in 28% of children examined by CTM because of dysraphism.[8] Diastematomyelia occurs in 5% of patients with congenital scoliosis and should be excluded in this clinical setting.[3] Diastematomyelia is usually discovered in children and is rare in adults.[2] It is frequently located in the lumbar region and is rare in the cervical spine.

Once diastematomyelia is suspected clinically, a conventional radiograph of the entire spine in AP and lateral projections frequently demonstrates spinal abnormalities, the most common being anomalies of the laminae and pedicles, which occur in approximately 90% of cases.[4] Spina bifida, vertical fusion of the laminae, and focal widening of the interpedicular distance are good predictors of the site of cord splitting.[4] A calcified spur, although visible in only 30% of cases,[9] is diagnostic of this disorder and also locates the site of cord splitting.[4]

CTM and conventional myelography can each demonstrate various features of diastematomyelia and other frequently associated disorders such as tethered conus, thickened filum terminale, myelomeningocele, hydromyelia, and intraspinal lesions such as lipomas and cysts.[1,6,9,10] Myelography with water-soluble contrast can be used as a primary examination to study the entire spinal canal for detection of diastematomyelia. This is particularly useful when the disorder is suspected clinically and its location is not evident radiographically. Myelography can also detect a second diastematomyelia.[5] However, CTM is superior to myelography in the detection of cord splitting, demonstration of the bony spur, and evaluation of associated skeletal abnormalities.[9] CTM can also salvage a suboptimal or difficult-to-interpret conventional myelogram. Thus, in cases where diastematomyelia is suspected, myelography and CTM can be used as complementary procedures. In the postoperative patient with recurrent symptoms, CTM can be used to demonstrate regrowth of a previously resected bony spur.[7]

References

1. Arredondo F, Haughton VM, Hemmy DC, et al: The computed tomographic appearance of the spinal cord in diastematomyelia. *Radiology* 1980;136:685–688.
2. Beyerl BD, Ojemann RG, Davis KR, et al: Cervical diastematomyelia presenting in adulthood. *J Neurosurg* 1985;62:449–453.
3. Giordano GB, Cerisoli M: Diastematomyelia and scoliosis. Usefulness of CT examination. *Spine* 1983;8:111–112.
4. Hilal SK, Marton D, Pollack E: Diastematomyelia in children. *Radiology* 1974;112:609–621.

5. McClelland RR, Marsh DG: Double diastematomyelia. *Radiology* 1977;123:378.
6. Naidich TP, Harwood-Nash DC: Diastematomyelia: Hemicord and meningeal sheaths; single and double arachnoid and dural tubes. *AJNR* 1983;4:633–636.
7. Pang D, Parrish RG: Regrowth of diastematomyelic bone spur after extradural resection: Case report. *J Neurosurg* 1983;59:887–890.
8. Pettersson H, Harwood-Nash DCF: *CT and Myelography of the Spine and Cord: Techniques, Anatomy and Pathology in Children*. Berlin, Springer-Verlag, 1982.
9. Scotti G, Musgrave MA, Harwood-Nash DC, et al: Diastematomyelia in children: Metrizamide and CT metrizamide myelography. *AJR* 1980;135:1225–1232.
10. Weinstein MA, Rothner AD, Duchesneau P, et al: Computed tomography in diastematomyelia. *Radiology* 1975;118:609–611.

CASE 43

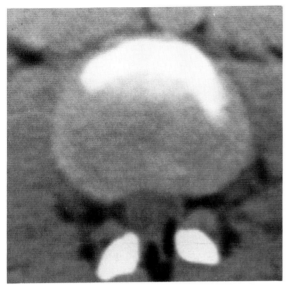

FIG. 43A. Axial CT scan at the level of the L2-L3 intervertebral disc in a 16-year-old female who had had spinal surgery at the age of 3 weeks for a benign teratoma.

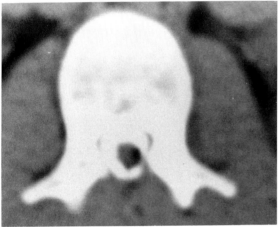

FIG. 43B. Axial CTM scan at L2 following introduction of intrathecal contrast.

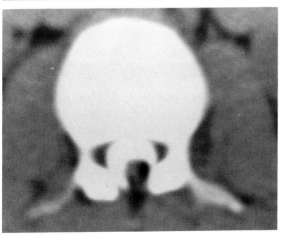

FIG. 43C. Axial CTM scan at the inferior aspect of L2, 8 mm caudad to Fig. 43B. Similar findings were evident on scans obtained at the L3 level.

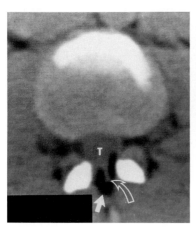 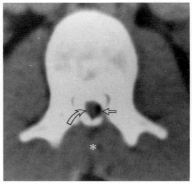 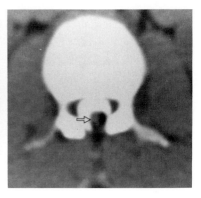

FIG. 43A-1. Tethered cord. There is a lobulated mass of low CT density (*straight arrow*) representing lipomatous tissue. The mass is adjacent to the thecal sac (*T*). A band of fibrosis (*curved arrow*) joins the thecal sac with the fatty mass. Although in this case the spinal cord cannot be identified without intrathecal contrast, tethered cord should be suspected.

FIG. 43B-1. CTM demonstrates the spinal cord (*curved arrow*) within the contrast-filled thecal sac. Lipomatous infiltration (*straight arrow*) is present adjacent to the cord. Spina bifida is noted (asterisk).

FIG. 43C-1. CTM at this level demonstrates lipomatous infiltration adjacent to a small cord (*arrow*).

Tethered Cord With Lipomatous Infiltration

The unenhanced CT scan demonstrates fatty infiltration and a fibrous band dorsal to the thecal sac (Fig. 43A-1). CTM reveals a low-lying spinal cord tethered by the lipomatous and fibrous mass (Figs. 43B-1, 43C-1). Additional scans demonstrated the cord to the L3 level. Spina bifida is also present. The diagnosis of tethered cord (conus) is made, and the extent of the lipomatous infiltration causing the tethering is established.

During gestation, the vertebral column develops at a more rapid rate than the spinal cord. Consequently, the conus medullaris "ascends" from the level of the coccyx to approximately the lower border of L2 at birth.[2] Two months after birth the conus is at its adult level, which is most often opposite L1 or L2 but may be anywhere from T12 to the L2-L3 intervertebral disc[2] (Fig. 43D). The diagnosis of tethered cord is established if the spinal cord lies below the level of the L2-L3 intervertebral disc after the age of 5 years.[3] In patients with tethered cord, the spinal cord is usually fixed (tethered) by one or more abnormalities such as a short, thickened filum terminale, an intradural lipoma, lipomatous infiltration, or fibrous adhesions.[7] The cord may also be tethered by the septum in diastematomyelia, by the neural plaque of myelomeningocele, or by adhesions that form at the site of a myelomeningocele repair.[4,7] The tethering anomalies cause compression or longitudinal traction on the spinal cord,[5,8] and traction may lead to cord ischemia, a possible cause of symptoms.[3]

The tethered cord syndrome usually presents during childhood, but milder cases may go undetected until the adult years.[1,5,7] Clinically there may be unexplained spastic gait, lower extremity weakness, scoliosis, foot deformity, or neurogenic bladder or bowel dysfunction.[3] Cutaneous manifestations of spinal dysraphism are frequent.[3,7]

Conventional radiographs reveal varying degrees of spina bifida, usually mild, involving at least one level in all cases.[4,7] A widened spinal canal and posterior scalloping may be present[5] but are not frequent. Both conventional myelography and CTM can demonstrate the presence of cord tethering, thickened filum terminale, intradural mass,[3,5] and other causes of cord tethering such as diastematomyelia, myelomeningocele, and lipomyelomeningocele. The normal CTM appearance of the adult lumbar spinal cord, lumbar enlargement, conus medullaris, and filum terminale are shown in Figs. 43E–43H. The diagnosis of tethered cord is made with CTM when the cord is seen below the level of L2-L3. Normally the filum terminale measures no more than 2 mm in diameter. A measurement greater than this is considered abnormal.[3,8] A low-lying thin cord can be differentiated from a thickened filum terminale if nerve roots are seen arising from the cord. With CTM, intradural lipoma and lipomatous infiltration of the filum or conus can be specifically diagnosed by the negative attenuation coefficient of fat tissue (Fig. 43I). CTM can define the junction between normal cord and lipomatous infiltration; this distinction cannot always be made with myelography.[5] Magnetic resonance imaging is another excellent modality for the evaluation of suspected tethered cord. An associated lipoma can be differentiated from other intraspinal masses by its high-intensity signal[6] (Fig. 43J).

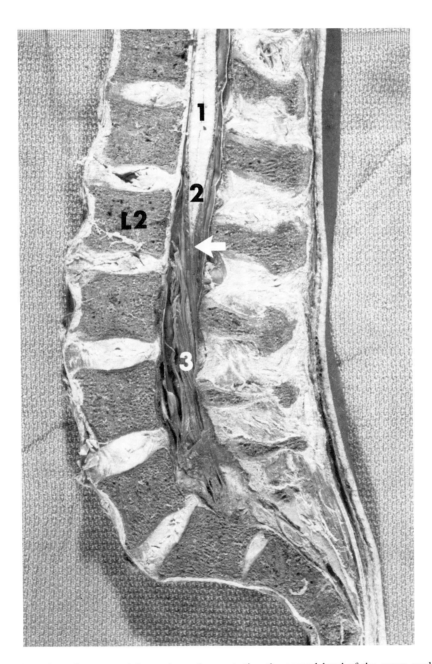

FIG. 43D. Sagittal section of a gross adult specimen demonstrating the normal level of the conus medullaris. *1*, spinal cord; *2*, conus medullaris; *3*, cauda equina; *arrow*, filum terminale.

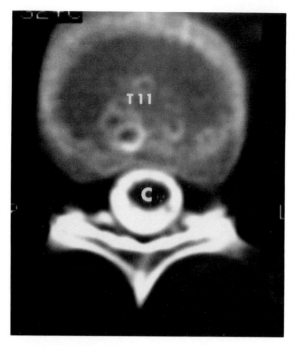

FIG. 43E. Normal spinal cord. CTM at T11 demonstrates a normal spinal cord (C) within the contrast-filled thecal sac.

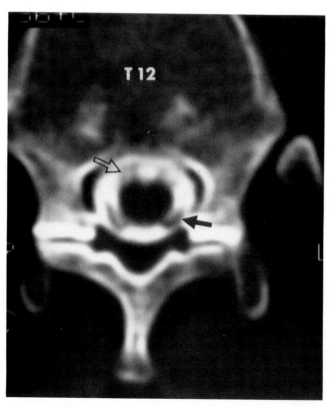

FIG. 43F. Normal lumbar plexus. CTM at T12. The spinal cord widens normally at the lumbar plexus. Ventral (*open arrow*) and dorsal (*closed arrow*) nerve roots are identified.

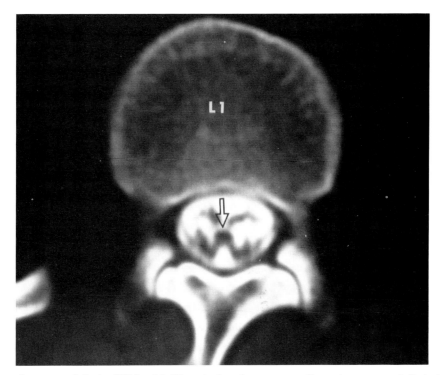

FIG. 43G. Normal conus medullaris. CTM at L1. The spinal cord tapers normally as the conus medullaris (*arrow*). Ventral and dorsal nerve roots arise from the conus and form the spiderlike configuration of the cauda equina.

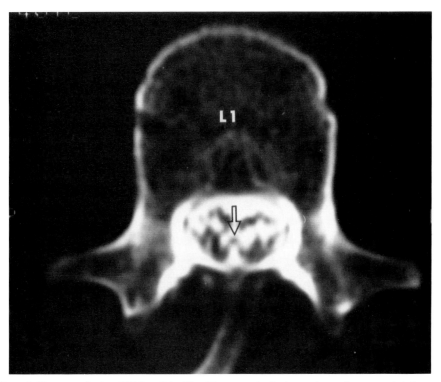

FIG. 43H. Normal filum terminale. CTM at L1. The filum terminale (*arrow*) continues caudad from the conus. It is normally no larger than 2 mm in diameter. The cauda equina is again identified by its spiderlike configuration.

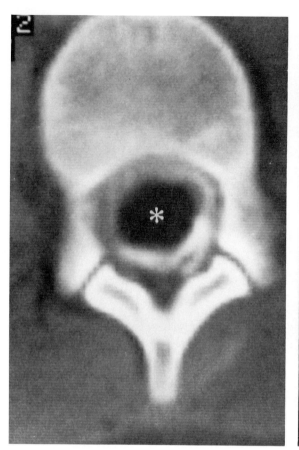 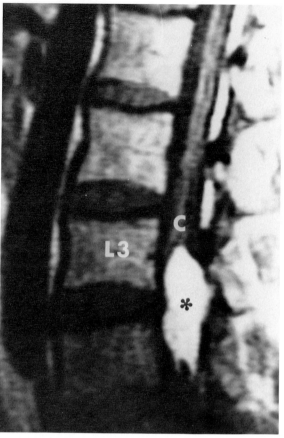

FIG. 43I. Intradural lipoma. Axial CTM at L3. There is a lipoma of negative CT attenuation value (asterisk). The contrast-filled subarachnoid space is compressed by the intradural mass. (Figure courtesy of Spencer Borden IV, M.D., Philadelphia, Pa.)

FIG. 43J. Tethered cord with lipoma. T1-weighted MRI of the lumbar region in the sagittal plane. The low-lying spinal cord (C) is tethered by a lipoma, which has bright intensity signals (asterisk).

References

1. Balagura S: Late neurological dysfunction in adult lumbosacral lipoma with tethered cord. *Neurosurgery* 1984;15:724–726.
2. Barson AJ: The vertebral level of termination of the spinal cord during normal and abnormal development. *J Anat* 1970;106: 489–497.
3. Fitz CR, Harwood-Nash DC: The tethered conus. *AJR* 1975;125:515–523.
4. Heinz ER, Rosenbaum AE, Scarff TB, et al: Tethered spinal cord following meningomyelocele repair. *Radiology* 1979;131: 153–160.
5. Kaplan JO, Quencer RM: The occult tethered conus syndrome in the adult. *Radiology* 1980;137:387–391.
6. Modic MT, Weinstein MA, Pavlicek W, et al: Nuclear magnetic resonance imaging of the spine. *Radiology* 1983;148: 757–762.
7. Pang D, Wilberger JE Jr: Tethered cord syndrome in adults. *J Neurosurg* 1982;57:32–47.
8. Sarwar M, Virapongse C, Bhimani S: Primary tethered cord syndrome: A new hypothesis of its origin. *AJNR* 1984;5:235–242.

CASE 44

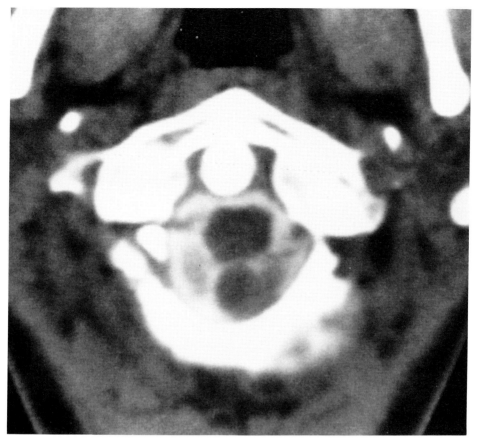

FIG. 44A. Axial CTM through the plane of C1 and the odontoid process anteriorly and the inferior aspect of the occiput posteriorly. This 23-year-old patient had a history of myelomeningocele repair at birth and was subsequently diagnosed as having hydrocephalus. She now presented with severe headaches and weakness of the upper extremities.

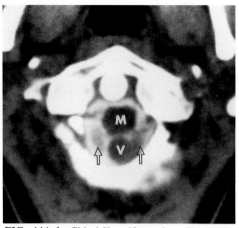

FIG. 44A-1. Chiari II malformation. The cerebellar vermis (*V*) is dorsal to the medulla (*M*), forming a figure-eight configuration. Cerebellar tonsils are noted posterolateral to the medulla (*arrows*). The medulla is larger, less rounded, and more irregular in the axial plane than the spinal cord or the vermis.

Chiari Malformation

The CTM examination demonstrates the medulla, cerebellar vermis, and cerebellar tonsils lying caudad to the foramen magnum, indicating a Chiari II malformation (Fig. 44A-1). The Chiari malformation is a congenital anomaly involving protrusion of the hindbrain into the cervical canal. In the Chiari I malformation, only cerebellar tonsils protrude through the foramen magnum.[4] The tonsils lie dorsal and lateral to the cervical cord and usually extend to the level of C1 or C2, whereas the medulla and fourth ventricle are in their normal positions.[3] Hydromyelia occurs frequently with Chiari I malformation, whereas hydrocephalus is less common. Severe spinal abnormalities and myelomeningocele are not associated with Chiari I malformation.[4] Chiari II malformation is more extensive and severe. There are varying degrees of downward protrusion of the medulla, pons, fourth ventricle, and cerebellar vermis through the foramen magnum into the cervical canal.[3] The medulla may displace the cervical cord inferiorly or may lie posterior to the cord. When pronounced, protrusion of the medulla causes a kink at the medulla-cord junction.[4] Herniation of small cerebellar hemispheres and tonsils is variable. The relatively large cerebellar vermis may lie dorsal to the medulla. Compression of the cervical cord or protruding structures may develop.[2] Patients with Chiari II malformation almost invariably are children with a myelomeningocele, and they may have associated hydromyelia or diastematomyelia.[2]

Conventional radiography may demonstrate a widened foramen magnum, small posterior fossa, wide upper cervical spinal canal, and spina bifida.[2] Conventional myelography and gas myelography can usually define Chiari malformation. The level of tonsil herniation discovered at surgery is lower than that suggested by myelography in over one third of cases of Chiari I malformation.[5] CTM is more accurate than conventional myelography or unenhanced CT in establishing the diagnosis.[1] CTM is useful in evaluating the nature and position of the protruding structures, the relationship of the protruding structures to the spinal cord, and the presence and degree of compression of neural tissue.[2,7]

In Chiari I malformation, protruding elongated cerebellar tonsils appear on CTM as crescent-shaped filling defects in the subarachnoid space posterolateral to the spinal cord[3] (Figs. 44B, 44C). They are usually asymmetric in shape since they often protrude to different levels. In Chiari II malformation, the CTM appearance depends on the nature and extent of the protruding structures. Cerebellar hemispheres appear as large posterolateral masses that may compress the medulla at or just below the foramen magnum. The overlapping of the cerebellar vermis dorsal to the medulla and the overlapping of the medulla dorsal to the cervical cord may demonstrate a figure-eight configuration[2] (Figs. 44A-1, 44B). The spinal cord may be compressed and flattened by the medulla. Sagittal reconstruction often helps demonstrate the caudal extent of the herniating structures. Hydromyelia, myelomeningocele, and other forms of dysraphism can also be evaluated. Additionally, the brain can be scanned for evaluation of hydrocephalus.

Magnetic resonance imaging can be used to evaluate the spinal cord in the sagittal and coronal planes and is an excellent modality for establishing a diagnosis of Chiari malformation[6] (Figs. 44D, 44E).

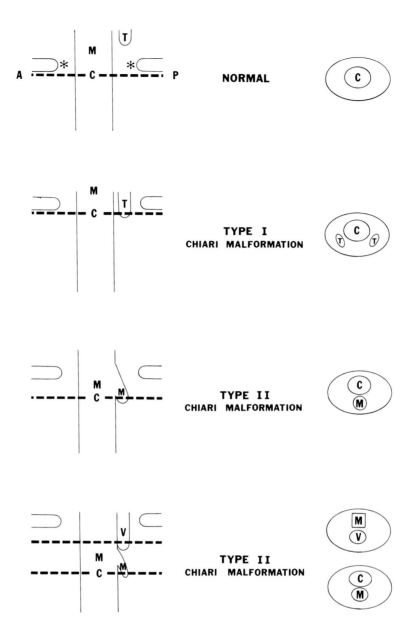

FIG. 44B. Some forms of Chiari malformations demonstrating the spatial relationships of the medulla *(M)*, cord *(C)*, cerebellar tonsils *(T)*, cerebellar vermis *(V)*, and foramen magnum *(asterisk)* in the sagittal and axial planes. *A*, anterior; *P*, posterior.

Normally, the junction of the medulla and the spinal cord is at the foramen magnum. The cerebellar structures are not inferior to the foramen magnum.

In Chiari I malformation, the medulla and spinal cord enjoy a normal relationship. However, the cerebellar tonsils *(T)* are caudad to the foramen magnum.

In Chiari II malformation, the medulla is caudad to the foramen magnum and forms a small kink with the cord. In the axial plane a figure-eight configuration may be identified.

In more severe cases of Chiari II malformation, the cerebellar vermis *(V)* is seen cascading on the medulla *(M)*, which is cascading on the cord *(C)*. The figure-eight configuration may be identified in the axial plane. The structures that make up the figure-eight configuration vary depending on the level of scanning.

The spatial relationships of the medulla, spinal cord, cerebellum, and fourth ventricle are easily recognized in the sagittal and coronal planes.

References

1. Forbes WStC, Isherwood I: Computed tomography in syringomyelia and the associated Arnold-Chiari type I malformation. *Neuroradiology* 1978;15:73–78.
2. Naidich TP, McLone DG, Fulling KH: The Chiari II malformation: Part IV: The hindbrain deformity. *Neuroradiology* 1983;25:179–197.
3. Naidich TP, McLone DG, Harwood-Nash DC: Malformations of the craniocervical junction, in Newton TH, Potts DG (eds): *Computed Tomography of the Spine and Spinal Cord.* San Anselmo, Calif, Clavadel Press, 1983, pp 355–366.
4. Northfield DWC: *The Surgery of the Central Nervous System.* London: Blackwell Scientific Publications, 1973.
5. Paul KS, Lye RH, Strang FA, et al: Arnold-Chiari malformation: Review of 71 cases. *J Neurosurg* 1983;58:183–187.
6. Spinos E, Laster DW, Moody DM, et al: MR evaluation of Chiari I malformations at 0.15 T. *AJNR* 1985;6:203–208. *AJR* 1985;144:1143–1148.
7. Woosley RE, Whaley RA: Use of metrizamide in computerized tomography to diagnose the Chiari I malformation. *J Neurosurg* 1982;56:373–376.

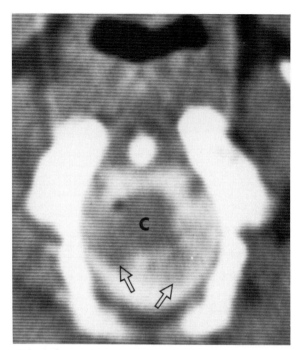

FIG. 44C. Chiari I malformation. Axial CTM at the level of C1. The cerebellar tonsils (*arrows*) are more caudad than normal as they lie posterolateral to the spinal cord (*C*).

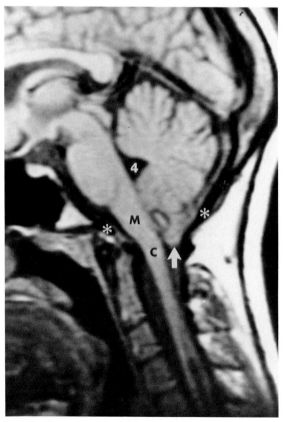

FIG. 44D. Chiari I malformation. T1-weighted MRI of the spinal cord (*C*) in the sagittal plane. The cerebellar tonsil (*arrow*) is caudad to the level of the foramen magnum (*asterisk*). *M*, medulla; *4*, fourth ventricle.

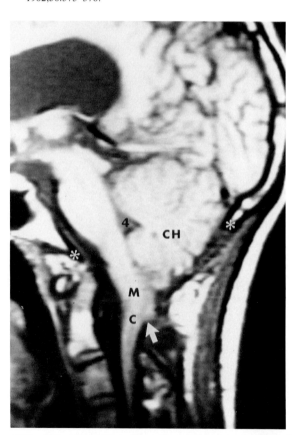

FIG. 44E. Chiari II malformation. T1-weighted MRI of the spinal cord (*C*) in the sagittal plane. The medulla (*M*) lies below the foramen magnum (*asterisk*) and is kinked (*arrow*) on the spinal cord. The cerebellar hemisphere (*CH*) and fourth ventricle (*4*) are more caudad than normal. Same patient as in Fig. 44A-1.

CASE 45

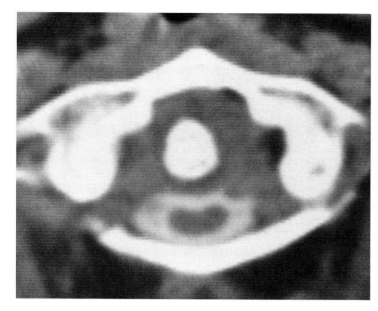

FIG. 45A. Axial CT at the C1-C2 level after intrathecal introduction of contrast. This 47-year-old male had a chronic illness and a 3-week history of numbness and weakness of the extremities.

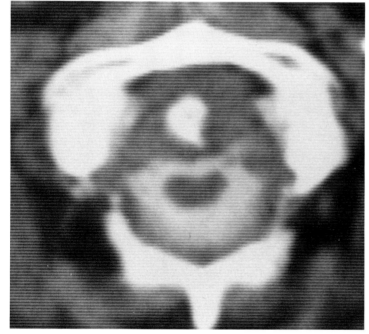

FIG. 45B. CTM 4 mm cephalad to Fig. 45A.

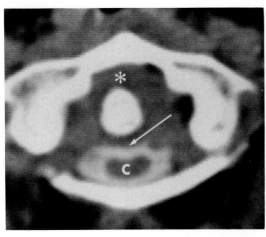 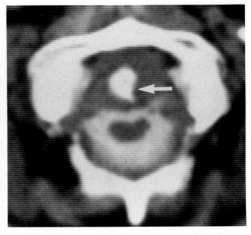

FIG. 45A-1. Rheumatoid arthritis with C1-C2 subluxation. There is widening of the space between the anterior arch of C1 and the odontoid process of C2 (*asterisk*). This distance measures 9 mm (normal distance is less than 3 mm in adults). The anterior C1-C2 subluxation is causing compression and displacement of the anterior subarachnoid space (*arrow*) and the spinal cord (*C*).

FIG. 45B-1. Erosion of the odontoid process is present (*arrow*). The anterior subluxation of C1-C2 can once again be seen causing spinal cord compression.

Rheumatoid Arthritis With C1-C2 Subluxation and Spinal Cord Compression

There is a 9-mm anterior subluxation at C1-C2 with spinal cord compression in this patient with rheumatoid arthritis (Figs. 45A-1, 45B-1). Normally, the strong transverse ligament helps prevent posterior displacement of the odontoid process. This ligament arches posterior to the odontoid process and attaches to tubercles on the medial wall of the lateral masses of C1 (Figs. 45C, 45D). Anterior subluxation occurs as a result of laxity or rupture of the transverse ligament due to the rheumatoid process. At the C1-C2 level, anterior subluxation is the most frequent type of subluxation occurring in patients with rheumatoid arthritis. It is diagnosed when the distance between the anterior arch of C1 and the odontoid process exceeds 3 mm in an adult or 5 mm in a child.[9] This measurement can readily be obtained by CT and is most accurately studied at bone window settings. Some patients with rheumatoid arthritis have lateral, posterior, or superior subluxation at C1-C2, which can also be evaluated by CT.

Another important anatomic consideration is the presence of two synovial membrane-lined joints about the odontoid process. The smaller joint lies between the anterior arch of C1 and the odontoid, whereas the larger posterolateral joint is located between the odontoid and the transverse ligament (Fig. 45C). Erosion of the odontoid process is a frequent finding in rheumatoid arthritis and occurs as a result of synovial inflammation (Fig. 45B-1). These erosions are more readily demonstrated with CT than with conventional radiography.[1,2]

The spinal cord is well demonstrated by CT in the upper cervical spine because it is surrounded by low-density cerebrospinal fluid in a large subarachnoid space. Spinal cord compression is diagnosed when there is posterior displacement of the cord or sufficient compression to cause alteration of its normal elliptical configuration. Another CT finding described in some patients with rheumatoid arthritis is a low-attenuation zone between the odontoid and the thickened transverse ligament that is thought to represent edema or inflammatory change in the synovial cavity.[2] Rarely a pathologic fracture of the odontoid process may be seen.

Many patients with rheumatoid arthritis are asymptomatic despite abnormalities of the craniocervical junction. Nevertheless, spinal cord compression with myelopathy may occur and is usually secondary to anterior C1-C2 subluxation exceeding 9 mm or to cranial settling (vertical subluxation).[11] Myelopathy is less common in the middle and lower cervical spine but may be caused by marked subluxation or abundant granulation tissue in the posterior portion of the spinal canal encroaching on the spinal cord.[4]

Besides rheumatoid arthritis, other causes of anterior subluxation at C1-C2 include psoriatic arthritis, juvenile rheumatoid arthritis and, less commonly, ankylosing spondylitis,[10] Marfan's syn-

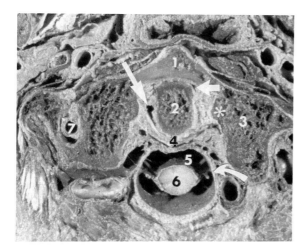

FIG. 45C. Normal anatomic specimen, transaxial section. The anterior arch of C1 (*1*) is in close approximation to the odontoid process of C2 (*2*). The lateral masses of C1 (*3*) have tubercles along their medial wall (*asterisk*), which act as attachments for the transverse ligament (*4*). Two synovial membrane-lined joints are located about the odontoid process and are partially visualized on this section. The larger joint lies posterolaterally between the odontoid and the transverse ligament (*long straight arrow*). The smaller anterior joint is found between the anterior arch of C1 and the odontoid (*short straight arrow*). This section displays other pertinent anatomy such as the dura (*curved arrow*), subarachnoid space (*5*), spinal cord (*6*), and vertebral artery (*7*) within the foramen transversarium.

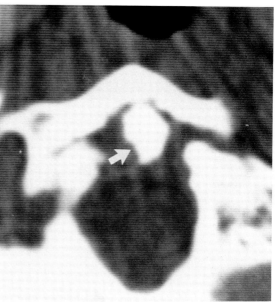

FIG. 45E. Multicentric reticulohistiocytosis. Axial CT scan at the level of C1 and the odontoid process. Erosion of the odontoid process (*arrow*) is more readily appreciated with CT than with conventional radiography. The posterior surface of the anterior arch of C1 is irregular and probably involved in the erosive process. The diagnosis of multicentric reticulohistiocytosis had been established 9 months earlier. The patient now presented with a 1-week history of increasing posterior cervical pain and numbness of the left occiput.

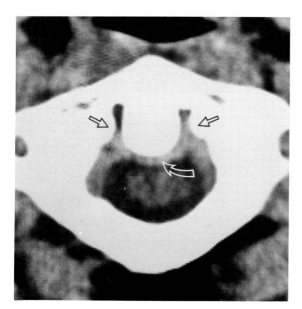

FIG. 45D. Normal transverse ligament. CT demonstrates the transverse ligament (*curved arrow*) arching posterior to the odontoid process and attaching to tubercles on the medial aspect of the lateral masses of C1 (*straight arrows*).

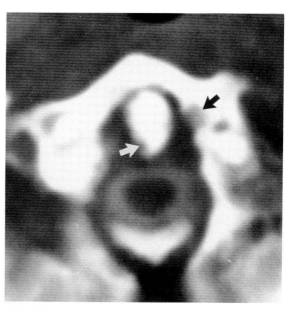

FIG. 45F. Multicentric reticulohistiocytosis. Same patient as in Fig. 45E. This scan was performed after intrathecal introduction of water-soluble contrast. Erosions of the odontoid (*white arrow*) and the lateral mass of C1 (*black arrow*) are present. The intrathecal contrast is not well seen because this scan was photographed at bone window settings; however, there is no evidence of spinal cord compression.

drome,[6] and Down's syndrome.[3] C1-C2 subluxation may also occur following trauma. Erosions, which are well defined by CT in rheumatoid arthritis, are also seen in the rheumatoid variants. Multicentric reticulohistiocytosis is an uncommon cause of erosive disease that may, like rheumatoid arthritis, involve the odontoid process, C1, and the occipital condyles[7] (Figs. 45E, 45F). The erosive changes may be very severe and, as in rheumatoid arthritis, may lead to cranial settling.[7]

CT is capable of demonstrating the anatomic detail needed for diagnosing various disorders of the craniocervical junction in patients with rheumatoid arthritis, its variants, and other rare disorders. Transaxial CT scans are obtained parallel to a line connecting the anterior and posterior arches of C1. This permits accurate measurement of the C1-C2 relationship and allows evaluation of the position of the odontoid process in relation to the foramen magnum. Obtaining a CT scan with the neck in flexion may permit identification of C1-C2 subluxation and cord compression not otherwise seen.[6,8] CTM is recommended by some investigators as an excellent method for the evaluation of the spinal cord in patients with rheumatoid arthritis.[5,8] There is good correlation between the severity of clinical symptoms and the deformity and displacement of the spinal cord at the C1-C2 level as well as at lower cervical levels.[5] Sagittal reformation may occasionally be helpful, particularly in patients with posterior C1-C2 subluxation.[1]

Despite the anatomic detail seen by CT, conventional radiography, including flexion and extension views in the lateral projection, remains the screening procedure of choice in evaluating the craniocervical junction in patients with rheumatoid arthritis. Some authors believe that CT evaluation should be reserved for those patients in whom conventional radiography and tomography do not explain the clinical findings and those cases in which surgical intervention is anticipated.[1] Others believe that CTM is indicated when a patient with cervical rheumatoid arthritis has neurological symptoms, especially when anterior C1-C2 subluxation exceeds 8 mm, cranial settling is progressive, or subaxial subluxation is suspected.[5]

References

1. Braunstein EM, Weissman BN, Seltzer SE, et al: Computed tomography and conventional radiographs of the craniocervical region in rheumatoid arthritis: A comparison. *Arthritis Rheum* 1984;27:26–31.
2. Castor WR, Miller JDR, Russell AS, et al: Computed tomography of the craniocervical junction in rheumatoid arthritis. *J Comput Assist Tomogr* 1983;7:31–36.
3. Hungerford GD, Akkaraju V, Rawe SE, et al: Atlanto-occipital and atlanto-axial dislocations with spinal cord compression in Down's syndrome: A case report and review of the literature. *Br J Radiol* 1981;54:758–761.
4. Kudo H, Iwano K, Yoshizawa H: Cervical cord compression due to extradural granulation tissue in rheumatoid arthritis: A review of five cases. *J Bone Joint Surg Br* 1984;66-B:426–430.
5. Laasonen EM, Kankaanpää U, Paukku P, et al: Computed tomographic myelography (CTM) in atlanto-axial rheumatoid arthritis. *Neuroradiology* 1985;27:119–122.
6. Levander B, Mellstrom A, Grepe A: Atlantoaxial instability in Marfan's syndrome. Diagnosis and treatment: A case report. *Neuroradiology* 1981;21:43–46.
7. Martel W, Abell MR, Duff IF: Cervical spine involvement in lipoid dermato-arthritis. *Radiology* 1961;77:613–617.
8. Osborne D, Triolo P, Dubois P, et al: Assessment of craniocervical junction and atlantoaxial relation using metrizamide-enhanced CT in flexion and extension. *AJNR* 1983;4:843–845.
9. Park WM, O'Neill M, McCall IW: The radiology of rheumatoid involvement of the cervical spine. *Skeletal Radiol* 1979;4:1–7.
10. Sorin S, Askari A, Moskowitz RW: Atlantoaxial subluxation as a complication of early ankylosing spondylitis: Two case reports and a review of the literature. *Arthritis Rheum* 1979;22:273–276.
11. Weissman BNW, Aliabadi P, Weinfeld MS, et al: Prognostic features of atlantoaxial subluxation in rheumatoid arthritis patients. *Radiology* 1982;144:745–751.

CASE 46

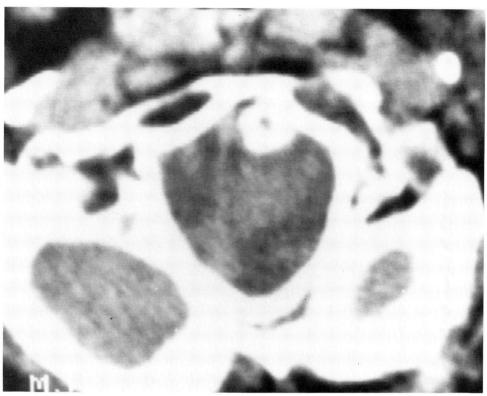

FIG. 46A. Axial CT at the level of the foramen magnum. This 59-year-old female has chronic systemic disease and severe neck pain.

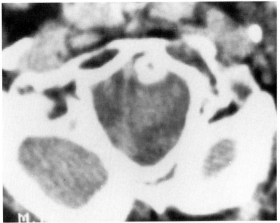

FIG. 46A-1. Cranial settling. The odontoid process extends above the foramen magnum. This patient had long-standing rheumatoid arthritis. She did not have a neurologic deficit at this time.

Cranial Settling in Rheumatoid Arthritis

The odontoid process is located above the level of the foramen magnum (Fig. 46A-1). This is the result of cranial settling in a patient with rheumatoid arthritis. Cranial settling is probably the most serious complication of rheumatoid arthritis. It occurs when pannus from the inflamed synovial joints leads to erosion and collapse of the lateral masses of C1 and to a lesser extent erosion of the occipital condyles and superior articular facets of C2.[3] These pathologic changes permit the skull to settle at a lower level on the cervical spine.

The diagnosis of cranial settling can be established with conventional radiography and tomography.[5,7] CT is indicated if confirmation of cranial settling is required or if there are signs of cord compression. The normal occipitocervical articulations are demonstrated in the coronal plane in Fig. 46B. When cranial settling is present, the odontoid process is seen above the level of the foramen magnum on a scan taken parallel to a line connecting the anterior and posterior arches of C1[1,2] (Figs. 46A-1, 46C). Compression of the upper spinal cord can frequently be determined without the need for intrathecal contrast. Erosive changes of the atlas and axis are much more common in patients with advanced cranial settling[6] and are readily demonstrated by CT (Figs. 46D, 46E).

Cranial settling occurs in 5% to 8% of patients with rheumatoid arthritis.[3] The most common symptom is occipital pain with radiation toward the skull vertex.[4] Neurologic symptoms such as paresthesias, paresis, or micturation disturbances occur in 30% of patients with rheumatoid arthritis and cranial settling.[6] There is a statistical correlation between the severity of vertical dislocation and the presence of neurologic symptoms.[6] In addition, patients with cranial settling are more likely to have neurologic symptoms when there is narrowing of the sagittal diameter of the canal at C1 to less than 13 mm.[6] Cranial settling typically occurs in patients with long-standing rheumatoid arthritis. Follow-up examination of these patients reveals progression of the vertical dislocation in 80% of cases.[6] Odontoid compression of the me-

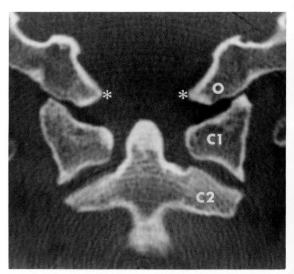

FIG. 46B. Normal occipitocervical articulations demonstrated in the coronal plane. The relationship between the occipital condyles (*O*), the lateral masses of C1, and the articular surfaces of C2 can be appreciated. The odontoid lies beneath the level of the foramen magnum (*asterisks*).

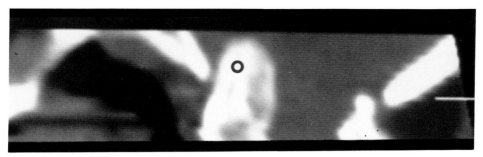

FIG. 46C. Cranial settling. Sagittal reconstruction of axial CT scans of a 56-year-old woman with a 16-year history of severe rheumatoid arthritis. She now presented with weakness of the upper and lower extremities and hyperreflexia of the lower extremities. The odontoid (*O*) is located above the level of the foramen magnum.

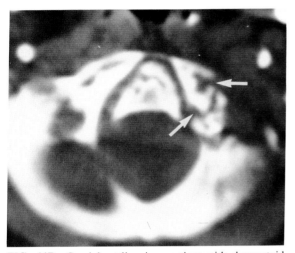

FIG. 46D. Cranial settling in a patient with rheumatoid arthritis. Same patient as in Fig. 46C. Axial CT through the base of the odontoid. This patient had extensive erosive disease with collapse of the lateral masses of C1 and the articular processes of C2 causing distortion of the normal anatomic appearance (*arrows*).

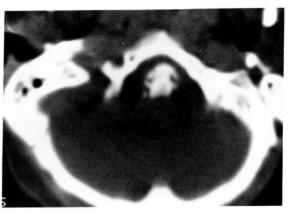

FIG. 46E. Cranial settling in a patient with rheumatoid arthritis. This scan, obtained cephalad to Fig. 46D, demonstrates erosions of the odontoid process.

dulla oblongata and spinal cord has led to death in some patients with cranial settling, and vertebral artery occlusion has been implicated as a cause of death in others.[3] Surgical intervention includes stabilization when cranial settling is reducible and decompression of the cervicomedullary junction when the abnormality is irreducible.[4]

References

1. Braunstein EM, Weissman BN, Seltzer SE, et al: Computed tomography and conventional radiographs of the craniocervical region in rheumatoid arthritis: A comparison. *Arthritis Rheum* 1984;27:26–31.
2. Castor WR, Miller JDR, Russell AS, et al: Computed tomography of the craniocervical junction in rheumatoid arthritis. *J Comput Assist Tomogr* 1983;7:31–36.
3. El-Khoury GY, Wener MH, Menezes AH, et al: Cranial settling in rheumatoid arthritis. *Radiology* 1980;137:637–642.
4. Menezes AH, VanGilder JC, Clark CR, et al: Odontoid upward migration in rheumatoid arthritis: An analysis of 45 patients with "cranial settling." *J Neurosurg* 1985;63:500-509.
5. Park WM, O'Neill M, McCall IW: The radiology of rheumatoid involvement of the cervical spine. *Skeletal Radiol* 1979;4:1–7.
6. Redlund-Johnell I, Pettersson H: Vertical dislocation of the C1 and C2 vertebrae in rheumatoid arthritis. *Acta Radiol Diagn* 1984;25:133–141.
7. Weissman BNW, Aliabadi P, Weinfeld MS, et al: Prognostic features of atlantoaxial subluxation in rheumatoid arthritis patients. *Radiology* 1982;144:745–751.

CASE 47

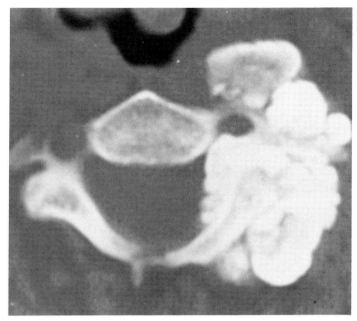

FIG. 47A. Axial CT at C2 in a 43-year-old woman with left neck pain as well as decreased motion and muscle weakness of the neck. In addition, she has muscle weakness of the right hand and arm thought to be related to calcifications present about the right shoulder.

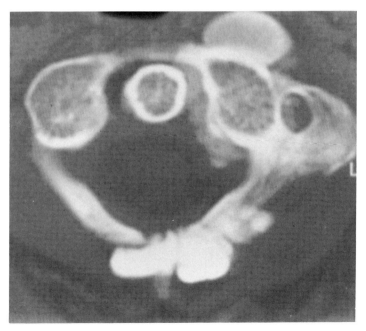

FIG. 47B. Axial CT through the plane of C1 and the odontoid process of C2.

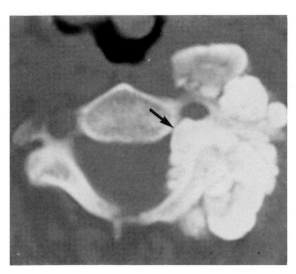 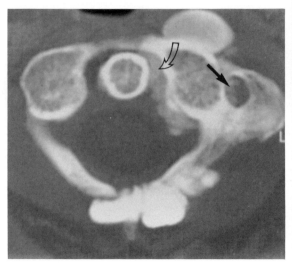

FIG. 47A-1. Scleroderma. There is massive amorphous calcification causing encroachment upon the spinal canal and neural canal (*arrow*).

FIG. 47B-1. Extensive calcification is causing narrowing of the foramen transversarium (*straight arrow*) and the space between C1 and the odontoid process (*curved arrow*).

Scleroderma with Massive Calcification

Massive intraspinal and paraspinal calcifications are present with extension into the neural foramen, the foramen transversarium, and the C1-C2 articulation (Figs. 47A-1, 47B-1). Additional CT sections revealed that this process extended from C1 to C3. These findings are due to dystrophic accumulation of hydroxyapatite crystals. The diagnosis of scleroderma was established in this patient approximately one year after the initial CT examination, as the clinical features of this disorder became apparent. This patient developed sclerodactyly and telangiectasia. Small focal calcifications in the subcutaneous tissues of the fingers were discovered radiographically.

The calcifications found in patients with scleroderma consist of hydroxyapatite crystals. These calcifications may develop within subcutaneous tissues, tendon sheaths, bursae and joints; and may appear punctate, amorphous, or linear.[4] Frequent sites of calcification include the hands, elbows, forearms, shoulders, and hips. Calcifications within the spine are rare.

In the case presented there are massive intraspinal and paraspinal calcifications. Although rare, a similar CT demonstration of massive calcific deposits at the craniovertebral junction has been reported in a patient with calcium pyrophosphate deposition disease.[1] In that case, spinal cord compression and osseous erosion were demonstrated by CT. Extensive heterotopic new bone formation may occur in the cervical spine in patients with fibrodysplasia ossificans progressiva (myositis ossificans progressiva)[3] and in patients who have sustained previous head injury.[2] CT can be useful in these cases to delineate the calcific or ossific masses earlier and with more anatomic detail than conventional radiography.[3] CT is helpful in treatment planning in those cases in which surgical intervention is contemplated.

References

1. El-Khoury GY, Tozzi JE, Clark CR, et al: Massive calcium pyrophosphate crystal deposition at the craniovertebral junction. *AJR* 1985;145:777-778.
2. Groswasser Z, Reider-Groswasser I: Heterotopic new bone formation in the cervical spine following head injury: Case report. *J Neurosurg* 1986;64:513-515.
3. Reinig JW, Hill SC, Fang M, et al: Fibrodysplasia ossificans progressiva: CT appearance. *Radiology* 1986;159:153-157.
4. Resnick D, Niwayama G: *Diagnosis of Bone and Joint Disorders: With Emphasis on Articular Abnormalities*, vol 2, Philadelphia, WB Saunders, 1981.

CASE 48

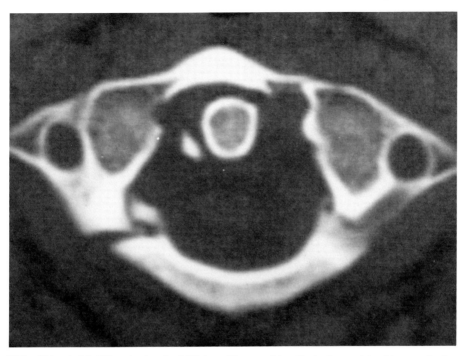

FIG. 48A. Axial CT at the level of C1 in a 20-year-old patient who sustained trauma to the cervical spine from a motor vehicle accident.

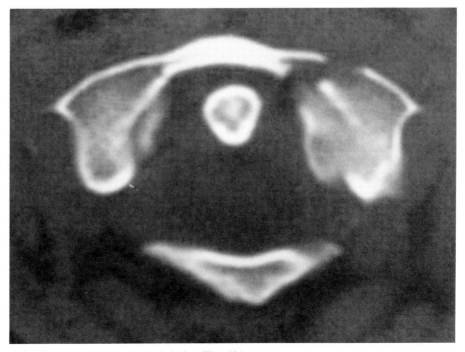

FIG. 48B. Axial CT 4 mm cephalad to Fig. 48A.

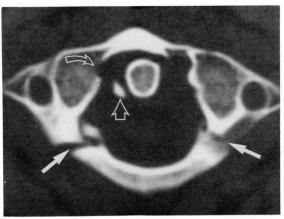
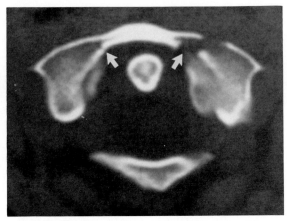

FIG. 48A-1. Jefferson fracture. There are bilateral fractures of the posterior arch of C1 (*closed arrows*) with a displaced fragment noted on the right. There is also a fracture of the medial tubercle of the lateral mass of C1 on the right (*curved arrow*) with displacement of a fragment into the canal (*straight open arrow*). The medial tubercle is the site of attachment of the transverse ligament. Widening of the space between the odontoid and the lateral masses of C1 can be appreciated.

Fig. 48B-1. There are bilateral fractures of the anterior arch of C1 (*arrows*). The fracture on the right is subtle and nondisplaced.

Jefferson Fracture

There are bilateral fractures of the posterior and anterior arches of C1 (Figs. 48A-1, 48B-1). A fracture of the medial tubercle of the lateral mass of C1 can also be seen. Coronal reconstruction of the axial images demonstrates lateral offset of the lateral masses of C1 in relation to the superior articular surfaces of C2 (Fig. 48C). This is a Jefferson fracture.

The Jefferson fracture is an injury that occurs from a direct blow to the vertex of the skull with the head held erect, resulting in compressive forces between the occiput and the articular surfaces of C2.[4,5] Because of the oblique plane of the articular surfaces, the transmitted force is directed laterally, leading to lateral spread of the lateral masses of C1.[5] This may in turn lead to tension fracture of the anterior and posterior arches of C1. The classic Jefferson fracture includes bilateral lateral offset of C1 and bilateral fractures of the posterior and anterior arches of C1. However, depending upon the degree of force and the position of the head at the time of injury, fractures may be unilateral or may be limited to the posterior or anterior arch.[5] Lateral offset is usually bilateral but may be unilateral.[4] Although some patients with Jefferson fracture die immediately at the time of injury, those patients presenting for clinical and radiographic evaluation of a Jefferson fracture usually do not have neurologic loss or compromise of the spinal cord.

The diagnosis of Jefferson fracture can be established by conventional radiography. Lateral offset of the lateral masses of C1 in relation to the superior articular surfaces of C2 is visualized with the open-mouth view. The fractures of the C1 ring usually occur near the junction of the arches and the lateral masses and are frequently not visualized with conventional radiography.[4] Conventional tomography may also fail to reveal the fractures or may be inconclusive.[7] CT is ideally suited for evaluating fractures of the atlas since the fractures through the ring of the atlas occur perpendicular to the plane of the CT scan and can thus be readily identified (Fig. 48D). The presence of a bone fragment within the spinal canal, one of the few indications for surgery in this injury, can be best demonstrated by CT.[6] Lateral displacement of the lateral masses of C1 may be visualized on the transaxial scans as well as with coronal reformation. When the lateral offset exceeds 7 mm, there is probably an associated tear of the transverse liga-

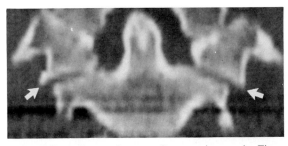

FIG. 48C. Jefferson fracture. Same patient as in Figs. 48A-1 and 48B-1. Coronal reconstruction of axial images demonstrates lateral offset of the lateral masses of C1 (*arrows*) in relation to the superior articular surfaces of C2.

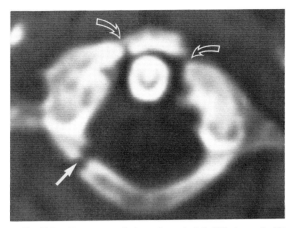

FIG. 48D. Fractures of the atlas. Axial CT through C1 demonstrates bilateral fractures of the anterior arch (*curved arrows*) and fracture of the right posterior arch (*straight arrow*).

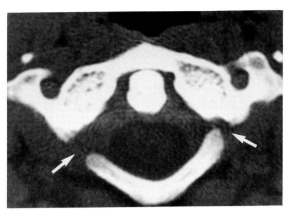

FIG. 48E. Posterior clefts of the atlas. Axial CT through C1 reveals bilateral posterior clefts (*arrows*). The smooth, rounded margins of the posterior arch help differentiate these congenital clefts from fractures, which typically have a more irregular, jagged appearance.

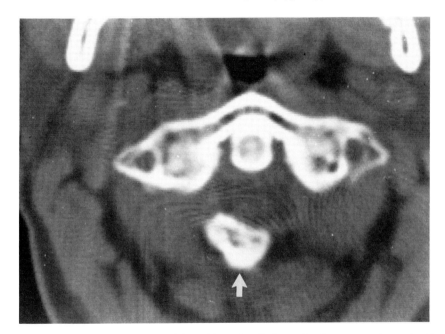

FIG. 48F. Partial aplasia of the posterior arch of C1. This patient had sustained trauma to the neck. Abnormal radiographs of the cervical spine led to this CT examination. There is partial aplasia of the posterior arch of C1 with a persistent posterior tubercle (*arrow*).

ment.[2] In the present case, fracture of the medial tubercle of the lateral mass of C1 is noted. This is the site of attachment of the transverse ligament.

The differential diagnosis of a Jefferson fracture includes congenital clefts, aplasia of the arch of the atlas, and pseudospread of the atlas. Congenital clefts are rare (Fig. 48E). Posterior clefts occur more frequently than anterior clefts and are most often in the midline.[3] Should lateral offset occur, it is limited to 1 to 2 mm.[3] Aplasia of the posterior arch of C1 may be partial or complete. Partial posterior aplasia may have a persistent posterior tubercle (Fig. 48F).[1] Clefts and aplasia can be differentiated from fracture by their smooth cortical margins.[1] Pseudospread of the atlas occurs in children aged 3 months to 4 years, most frequently in the second year.[8] Pseudospread results from more rapid growth of C1 in relation to C2 leading to apparent lateral offset of C1 as visualized on AP open-mouth radiographs.[8] In comparison, the Jefferson fracture is rare in children.

Fractures of the atlas comprise 5% of all cervical fractures and dislocations.[4] The most common

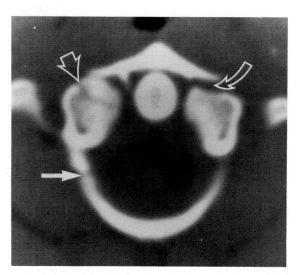

FIG. 48G. Fractures of the atlas. Bilateral fractures of the anterior arch of C1 (*curved arrow*) and fracture of the right posterior arch of C1 (*closed arrow*) are demonstrated. There is a fracture of the lateral mass of C1 on the right (*straight open arrow*).

fractures of the atlas are posterior arch fracture, Jefferson fracture, and horizontal fracture of the anterior arch.[4] Fracture of the lateral mass of C1 is usually associated with additional fractures of the atlas (Fig. 48G). Less common fractures of the atlas include fracture of the transverse process and isolated fracture of the lateral mass.

References

1. Dorne HL, Just N, Lander PH: CT recognition of anomalies of the posterior arch of the atlas vertebra: Differentiation from fracture. *AJNR* 1986;7:176-177.
2. Fielding JW, Cochran GVB, Lawsing JF, et al: Tears of the transverse ligament of the atlas: A clinical and biomechanical study. *J Bone Joint Surg Am* 1974;56-A:1683–1690.
3. Gehweiler JA Jr, Daffner RH, Roberts L Jr: Malformations of the atlas vertebra simulating the Jefferson fracture. *AJNR* 1983;4:187–190; *AJR* 1983;140:1083–1086.
4. Gehweiler JA Jr, Duff DE, Martinez S, et al: Fractures of the atlas vertebra. *Skeletal Radiol* 1976;1:97–102.
5. Jefferson G: Fracture of the atlas vertebra: Report of four cases, and a review of those previously recorded. *Br J Surg* 1920;7:407–422.
6. Kershner MS, Goodman GA, Perlmutter GS: Computed tomography in the diagnosis of an atlas fracture. *AJR* 1977;128:688–689.
7. Steppé R, Bellemans M, Boven F, et al: The value of computed tomography scanning in elusive fractures of the cervical spine. *Skeletal Radiol* 1981;6:175–178.
8. Suss RA, Zimmerman RD, Leeds NE: Pseudospread of the atlas: False sign of Jefferson fracture in young children. *AJNR* 1983;4:183–186; *AJR* 1983;140:1079–1082.

CASE 49

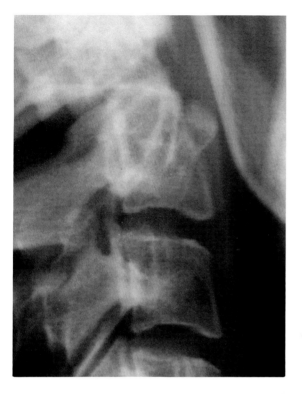

FIG. 49A. Lateral radiograph of the upper cervical spine. This patient had sustained trauma to the cervical spine. The remainder of the radiographic examination of the cervical spine was unremarkable.

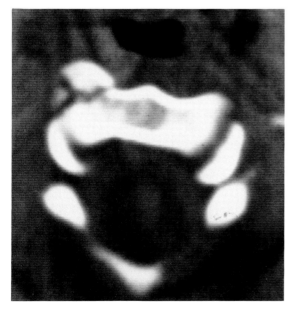

FIG. 49B. Axial CT scan at the level of the C2 vertebral body just caudad to the odontoid process.

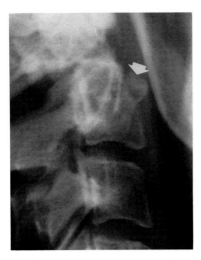
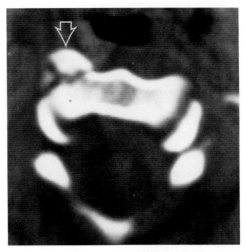

FIG. 49A-1. A suspicious bone density is seen anterior to the C2 vertebral body (*arrow*). There is no soft-tissue swelling.

FIG. 49B-1. Fracture of C2. A fracture on the right side of C2 is clearly demonstrated (*arrow*). The anteriorly displaced fracture fragment accounts for the presence of the bone density anterior to C2 as seen in Fig. 49A-1.

Fracture of C2

The lateral radiograph of the cervical spine demonstrates a bone density anterior to the C2 vertebral body (Fig. 49A-1). There is no soft-tissue swelling, and additional views failed to demonstrate other evidence of fracture. A CT scan was performed and clearly demonstrates a fracture on the right side of the C2 vertebral body (Fig. 49B-1). The anteriorly displaced fracture fragment identified with CT explains the suspicious bone density seen with conventional radiography.

The most frequent fracture of C2 is fracture of the odontoid process, which accounts for 13% of cervical spine fractures and/or dislocations.[6] These fractures may go undetected with conventional radiographic studies, and further evaluation with conventional tomography or CT may be needed. CT may also fail to demonstrate odontoid fracture, especially when the fracture is horizontal and nondisplaced.[2] In this setting the plane of scanning is parallel to the fracture line, and the nondisplaced fracture may not be visualized on the transaxial scan because of volume averaging. In some cases, sagittal and coronal reconstruction views may be the best or only way to identify these fractures with CT (Figs. 49C, 49D). Thin axial sections (1.5 or 2.0 mm) are needed to obtain best quality reconstruction. Axial CT and sagittal reconstruction views can help evaluate displacement of a fracture and the degree of spinal canal and spinal cord compromise (Fig. 49E, 49F). Patient cooperation, which is always important, is particularly necessary when studying odontoid fractures since patient movement significantly detracts from the reformatted image (Figs. 49G, 49H).

Odontoid fractures are classified into three types.[1] Avulsion fractures located at the tip of the odontoid process are of type I and are uncommon but stable fractures. The type II fracture occurs at the junction of the odontoid process and the body of C2. This is the most common type of odontoid fracture. It is unstable and has a high incidence of nonunion when treated conservatively.[1] Nonunion may develop whether the fracture is displaced or nondisplaced. Fractures that are initially nondisplaced frequently become displaced if fusion is not performed.[1] The type III fracture extends into the body of C2 and infrequently leads to nonunion. Fracture of the odontoid may be associated with Jefferson fracture of C1 or atlantoaxial subluxation and should be sought when these injuries are present[9] (Figs. 49G, 49H).

The second most common fracture of the axis is the so-called "hangman's fracture," with bilateral fractures of the pedicles of C2 occurring anterior to the inferior articular processes (Fig. 49I). This injury results from acute hyperextension of the skull upon the cervical spine and accounts for 7% of cases of fracture and/or dislocation of the cervical spine.[4] The bilateral fracture through the pedicles may or may not be associated with anterior displacement of the C2 vertebral body on C3 (traumatic spondylolisthesis of the axis). Anterior slippage of C2 occurs as a result of disruption of the anterior longitudinal ligament. In addition there may be an avulsion fracture of the anterior inferior aspect of the C2 vertebral body or the anterior superior margin of C3.[3] Much less often there may be atypical fractures of the neural arch of C2 with fracture of both laminae of C2 (Fig. 49J) or fracture of one pedicle and the opposite lamina. These atypical neural arch fractures as well as the typical bilateral pedicular fractures are considered unstable but are usually not associated with significant neurologic deficit. They usually heal well with conservative management.[3-7] Bilateral symmetric fractures of the arch of C2 are usually readily apparent on the lateral radiograph of the cervical spine. However, fractures may be more difficult to detect when they occur asymmetrically at different sites of the vertebral arch

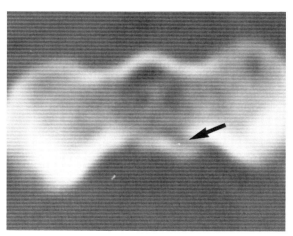

FIG. 49C. Fracture at the base of the odontoid process. Axial CT at the base of the odontoid process suggests a subtle fracture (*arrow*). This injury could not be further delineated by additional axial scans.

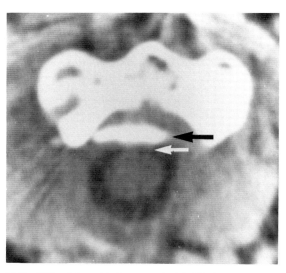

FIG. 49E. Odontoid fracture. Axial CT demonstrates a fracture at the base of the odontoid process. The posteriorly displaced fracture fragment (*black arrow*) abuts the anterior subarachnoid space (*white arrow*) without compressing or displacing the spinal cord.

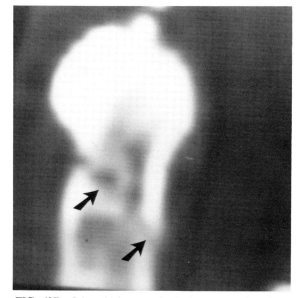

FIG. 49D. Odontoid fracture. Sagittal reconstruction. Same patient as in Fig. 49C. In this case, the use of sagittal reconstruction permitted better visualization of the fracture and its orientation. The fracture is not displaced and has an oblique course (*arrows*).

FIG. 49F. Odontoid fracture. Sagittal reconstruction. Same patient as in Fig. 49E. The oblique course of the fracture is apparent (*arrows*), with the fracture involving both the odontoid and the body of C2 (type III odontoid fracture).

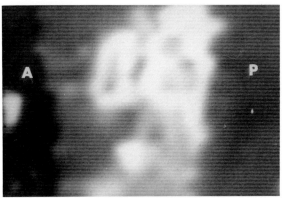

FIG. 49G. Unsatisfactory sagittal reconstruction due to inability of the patient to cooperate. A Jefferson fracture of C1 was diagnosed with axial CT in this uncooperative trauma victim. Considerable patient motion detracted from the image quality and prevented adequate reconstruction of the axial images. Although no fracture of the odontoid process could be seen, a repeat examination was recommended. *A*, anterior; *P*, posterior.

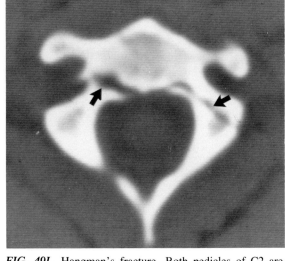

FIG. 49I. Hangman's fracture. Both pedicles of C2 are fractured (*arrows*). This type of injury may be associated with traumatic spondylolisthesis of the axis.

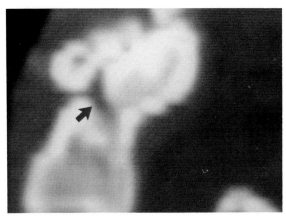

FIG. 49H. Fracture of the odontoid process in a patient with Jefferson fracture. Sagittal reconstruction. Same patient as in Fig. 49G. A repeat examination was obtained when the patient was able to cooperate. An oblique angulated fracture of the odontoid could now be seen (*arrow*). This case demonstrates the need for patient cooperation in obtaining diagnostic reconstruction views.

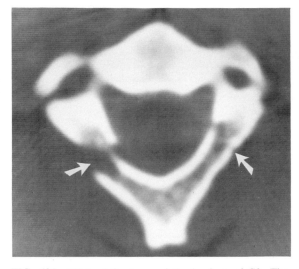

FIG. 49J. Bilateral fractures of the laminae of C2. The posterior elements are well visualized with CT. In this case fractures of the laminae can be seen (*arrows*).

(and are thus obscured by superimposition of solid bone from the opposite side).[8] In questionable cases, CT can readily identify these fractures.

References

1. Anderson LD, D'Alonzo RT: Fractures of the odontoid process of the axis. *J Bone Joint Surg* 1974;56-A:1663–1674.
2. Brant-Zawadzki M, Miller EM, Federle MP: CT in the evaluation of spine trauma. *AJR* 1981;136:369–375.
3. Elliott JM, Rogers LF, Wissinger JP, et al: The hangman's fracture: Fractures of the neural arch of the axis. *Radiology* 1972;104:303–307.
4. Gehweiler JA Jr, Clark WM, Schaaf RE, et al: Cervical spine trauma: The common combined conditions. *Radiology* 1979; 130:77–86.
5. Gehweiler JA Jr, Osborne RL, Becker RF. *The Radiology of Vertebral Trauma.* Philadelphia, WB Saunders, 1980.
6. Martinez S, Morgan CL, Gehweiler JA Jr, et al: Unusual fractures and dislocations of the axis vertebra. *Skeletal Radiol* 1979;3: 206–212.
7. Miller MD, Gehweiler JA Jr, Martinez S, et al: Significant new observations on cervical spine trauma. *AJR* 1978;130:659–663.
8. Rogers LF: *Radiology of Skeletal Trauma.* New York, Churchill Livingstone, 1982.
9. Sherk HH: Fractures of the atlas and odontoid process. *Orthop Clin North Am* 1978;9:973–984.

CASE 50

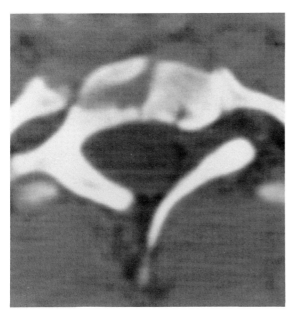

FIG. 50A. Axial CT at T1 obtained because of trauma to the lower cervical and upper thoracic spine.

FIG. 50B. Axial CT obtained 4 mm caudad to Fig. 50A.

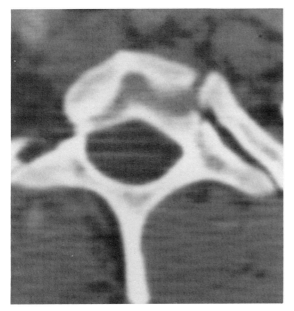

FIG. 50C. Axial CT obtained 2 mm caudad to Fig. 50B.

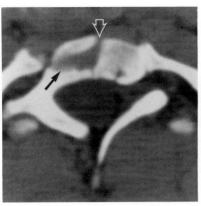 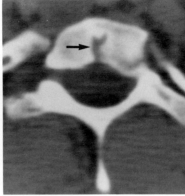 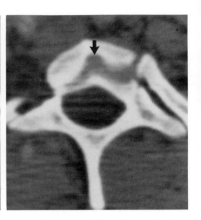

FIG. 50A-1. Butterfly vertebra. An irregular band of intervertebral disc (*black arrow*) is seen at this midvertebral body level. A cogenital cleft is noted anteriorly (*white arrow*).

FIG. 50B-1. The midline cleft (*arrow*) is clearly identified and has sclerotic margins. This cleft should not be confused with a fracture.

FIG. 50C-1. Immediately caudad to Fig. 50B-1, the inwardly protruding disc (*arrow*) is again demonstrated at the midvertebral body level. This sequence of scans demonstrates the typical axial CT appearance of butterfly vertebra.

Butterfly Vertebra

Butterfly vertebra is a developmental abnormality in which there is failure of fusion of the lateral halves of the vertebral body due to persistence of notochord tissue.[9] The vertebral surfaces become depressed centrally by disc tissue protruding into the cleft, thus causing a "butterfly" configuration when viewed on the AP radiograph. Butterfly vertebra is usually solitary but may be multiple in congenital syndromes such as dysraphism and may be associated with gastrointestinal, genitourinary, or central nervous system abnormalities.

The sagittal cleft of the butterfly vertebra is easily identified on the axial CT scan (Figs. 50A-1–50C-1). The cleft may have sclerotic margins and should not be confused with an acute fracture. Irregularly shaped bands of soft-tissue density representing inwardly protruding disc tissue may be seen both above and below the midportion of the vertebral body. This typical appearance should be diagnostic; however, coronal reconstruction of the axial images and conventional AP radiographs will resolve any problem cases (Figs. 50D, 50E).

Other developmental and acquired abnormalities of the spine may simulate fracture or a destructive process. For example, a limbus vertebra may be confused with a fracture of the vertebral body. Limbus vertebra occurs when disc tissue herniates anteriorly, penetrating vertebral trabeculae at the junction of the cartilaginous endplate and the bony vertebral rim.[8,9] This causes separation of small bony fragments from the vertebral body. Limbus vertebra occurs most frequently in the lumbar spine at the anterosuperior margin of the vertebral body. The lateral surface is less frequently involved.[8] Limbus vertebra is usually an incidental finding on the conventional radiograph or CT examination and has a characteristic appearance.

Conventional radiographs and CT of a limbus vertebra demonstrate a triangular or rounded, well-corticated bony fragment adjacent to the anterosuperior margin of the vertebral body (Figs. 50F, 50G). The fragment is separated from the vertebral body by a radiolucent band that represents herniated disc tissue. More than one bone fragment may be present. The posterior margin of the fragments and the anterior margin of the adjacent vertebral body appear sclerotic.[13] Intervertebral disc space height is usually normal or slightly decreased.

Another anomaly that may be encountered on CT is os odontoideum (Figs. 50H, 50I). Os odontoideum is a small, round or oval, corticated ossicle that is cranial to a hypoplastic odontoid process. The ossicle may lie adjacent to the tip of the odontoid or may be more cephalad near the basion. Most authors feel that os odontoideum is a congenital anomaly;[6,11] however, others believe that it occurs secondary to childhood trauma.[2,3] Associated osseous abnormalities of C1 include hypoplasia of the posterior arch and hypertrophy of the anterior arch.[11]

Os odontoideum is frequently associated with atlantoaxial instability. This instability depends on the level of the cleft in the odontoid process.[11] If the transverse ligament is in juxtaposition to the os odon-

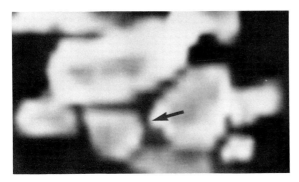

FIG. 50D. Butterfly vertebra. Same patient as in Fig. 50A-1. Coronal reconstruction of axial CT images demonstrates the butterfly vertebra of T1 with a midline cleft (arrow). Congenital abnormalities such as this may appear somewhat confusing on axial CT. Reconstruction of axial images and conventional radiography are helpful in understanding the complex axial images.

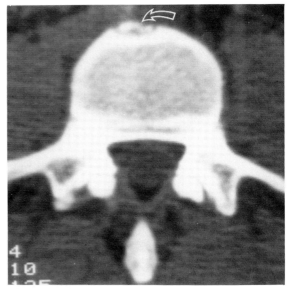

FIG 50F. Limbus vertebra. There is an irregular bone fragment seen anteriorly (arrow) separated from the vertebral body by a thin radiolucent band. Note sclerosis of the anterior margin of the vertebral body.

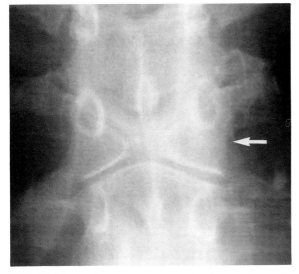

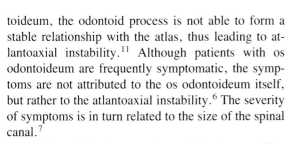

FIG. 50E. Butterfly vertebra. Conventional AP radiograph of another patient demonstrates butterfly vertebra of T4 (arrow).

toideum, the odontoid process is not able to form a stable relationship with the atlas, thus leading to atlantoaxial instability.[11] Although patients with os odontoideum are frequently symptomatic, the symptoms are not attributed to the os odontoideum itself, but rather to the atlantoaxial instability.[6] The severity of symptoms is in turn related to the size of the spinal canal.[7]

Congenital absence of a pedicle is rare (Fig. 50J). It is found in the lumbar, cervical, and thoracic spine in decreasing order of frequency.[5] Most cases of congenital absence of the pedicle are discovered

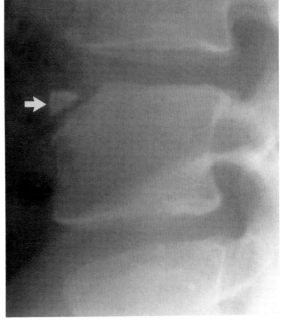

FIG. 50G. Limbus vertebra. Lateral radiograph of the lumbar spine demonstrates a limbus vertebra. The bone fragment (arrow) is derived from the anterior superior aspect of the vertebral body. The fragment is separated from the vertebral body by disc tissue, which has herniated anteriorly and inferiorly.

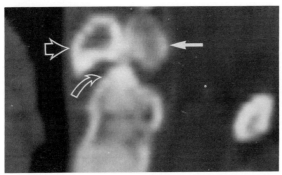

FIG. 50H. Os odontoideum. Sagittal reconstruction of axial CT images at C1 and C2 levels. An os odontoideum (*closed arrow*) is seen cephalad to a hypoplastic odontoid process (*curved arrow*). There is hypertrophy of the anterior arch of C1 (*straight open arrow*). In this case the anterior superior aspect of the odontoid lies anterior to the inferior portion of the os odontoideum. This sagittal reconstruction view is helpful in understanding the axial scan shown in Fig. 50I.

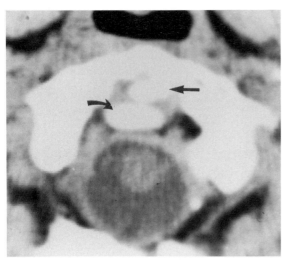

FIG. 50I. Os odontoideum. Axial CT image through the plane of the superior portion of the odontoid (*straight arrow*) and the inferior aspect of the os odontoideum (*curved arrow*). This somewhat confusing axial image is clarified by the sagittal reconstruction shown in Fig. 50H.

fortuitously, and the patient's signs and symptoms are rarely due to the absent pedicle.[12] However, pedicle defects may be associated with genitourinary and other congenital abnormalities[14] and rarely occur in neurofibromatosis.[12] Congenital absence of the pedicle should not be mistaken for destruction of a pedicle by tumor. With congenital absence (or hypoplasia) of the pedicle there are usually associated bony alterations. These include hypertrophy and sclerosis of the contralateral pedicle and vertebral arch due to added bone stress.[4,10] In addition, there is often hypoplasia or absence of the ipsilateral superior articular process and widening of the intervertebral foramen.[1,4,10,14] In

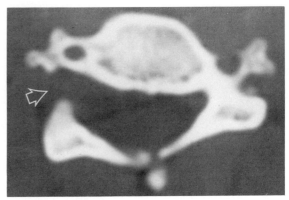

FIG. 50J. Congenital absence of the right pedicle of C4 (*arrow*) is demonstrated on this axial CT scan as well as on adjacent sections. This should not be confused with a destructive lesion of the pedicle.

comparison, destruction of the pedicle by tumor or infection is not associated with bony alterations of hypertrophy, sclerosis, or hypoplasia that are found with congenital absence of the pedicle. Instead, pedicle destruction is likely to be associated with a soft-tissue mass or additional destruction of adjacent bone.

References

1. Cox HE, Bennett WF: Computed tomography of absent cervical pedicle: Case report. *J Comput Assist Tomogr* 1984;8:537–539.
2. Fielding JW, Griffin PP: Os odontoideum: An acquired lesion. *J Bone Joint Surg Am* 1974;56-A:187–190.
3. Freiberger RH, Wilson PD, Nicholas JA: Acquired absence of the odontoid process. *J Bone Joint Surg Am* 1965;47-A:1231–1236.
4. Maldague BE, Malghem JJ: Unilateral arch hypertrophy with spinous process tilt: A sign of arch deficiency. *Radiology* 1976;121:567–574.
5. Manaster BJ, Norman A: CT diagnosis of thoracic pedicle aplasia: Case report. *J Comput Assist Tomogr* 1983;7:1090–1091.
6. McRae DL: The significance of abnormalities of the cervical spine. *AJR* 1960;84:3–25.
7. Minderhoud JM, Braakman R, Penning L: Os odontoideum: Clinical, radiological and therapeutic aspects. *J Neurol Sci* 1969;8:521–544.
8. Resnick D, Niwayama G: Intravertebral disk herniations: Cartilaginous (Schmorl's) nodes. *Radiology* 1978;126:57–65.
9. Schmorl G, Junghanns H: *The Human Spine in Health and Disease,* ed. 2. New York, Grune & Stratton, 1971.
10. Tomsick TA, Lebowitz ME, Campbell C: The congenital absence of pedicles in the thoracic spine: Report of two cases. *Radiology* 1974;111:587–589.
11. Wollin DG: The os odontoideum. *J Bone Joint Surg Am* 1963; 45-A:1459–1471,1484.
12. Wortzman G, Steinhardt MI: Congenitally absent lumbar pedicle: A reappraisal. *Radiology* 1984;152:713–718.
13. Yagen R: CT diagnosis of limbus vertebra: Case report. *J Comput Assist Tomogr* 1984;8:149–151.
14. Yousefzadeh DK, El-Khoury GY, Lupetin AR: Congenital aplastic-hypoplastic lumbar pedicle in infants and young children. *Skeletal Radiol* 1982;7:259–265.

CASE 51

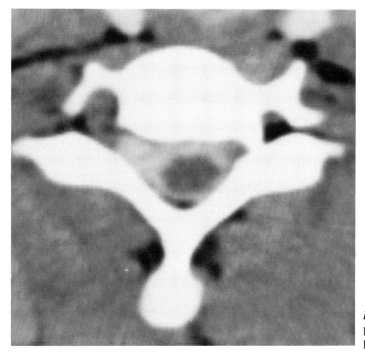

FIG. 51A. Axial CTM at C6-C7. This patient sustained a traumatic brachial palsy.

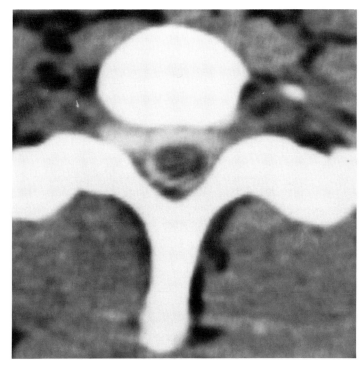

FIG. 51B. Axial CTM at C7-T1.

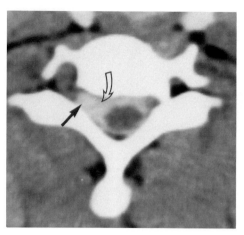 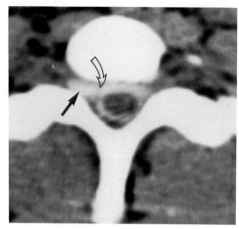

FIG. 51A-1. Nerve root avulsion. There is an outpouching of contrast (*straight arrow*) within the neural foramen. This collection of contrast represents a meningocele, which occurs from avulsion of the C7 nerve root. The meningocele is separated from the contrast-filled subarachnoid space by a thin dural plane (*curved arrow*). Slight displacement of the spinal cord is noted.

FIG. 51B-1. Avulsion of the C8 nerve root is diagnosed as contrast extends ventrolaterally into the neural foramen at the C7-T1 level (*straight arrow*). Again, a thin dural plane (*curved arrow*) separates the meningocele from the subarachnoid space.

Nerve Root Avulsion

In this case CTM demonstrates collections of contrast extending into the neural foramina at the level of the C7 and C8 nerve root sleeves (Figs. 51A-1, 51B-1). This is due to cervical nerve root avulsion. A cervical nerve root avulsion appears on CTM examination as an expanded outpouching of contrast extending ventrolaterally from the thecal sac into the neural foramen.[4,6] This collection of contrast, which represents a meningocele, is sometimes separated from the thecal sac by a thin dural plane measuring 1 to 2 mm in width.[6] At the time of injury the avulsed nerve root retracts and is usually not seen within the meningocele during CTM study.

Cervical nerve root avulsion may lead to a brachial plexopathy or palsy. The C7, C8, and T1 nerve roots are most commonly involved when the injury occurs while the arm is in abduction.[6] Traction is greatest on the C5 and C6 nerve roots when the arm is in adduction.[6] CTM and myelography are both imaging modalities that can demonstrate cervical nerve root avulsion. CTM offers the additional benefit of demonstrating the presence of bone fracture fragments and hematoma, which could be correctable causes of symptoms. CTM is also useful in diagnosing a dural tear with or without associated nerve root avulsion. The dural laceration leads to contrast escaping outside the subarachnoid space, a finding that is more readily visualized by CTM than by conventional myelography in some cases.[4]

Conventional CT without contrast is often used in evaluation of cervical trauma and is helpful in demonstrating the location and configuration of cervical fractures, displacement of fracture fragments, and degree of spinal cord compression (Fig. 51C). CT is particularly important in the evaluation of C6, C7, and T1 since these levels may be difficult to evaluate adequately by conventional radiography.[8] Radiographic evaluation of patients with cervical spine trauma can be accomplished with combined conventional radiography and CT, often without need for conventional tomography.[1]

The vertebral arch is the most frequent site of fracture in cervical spine injury. Fractures of the vertebral arch are found in 50% of patients with cervical spine fracture and/or dislocation.[3] From C3 to C7, the pillar fracture is the most frequent type of vertebral arch fracture.[3] The pillar is a rhomboid structure formed by the fusion of the superior and inferior articular process of each vertebra (e.g., the superior and inferior articular processes of C5 fuse to form the pillars of C5). Fractures through the pillars can be detected by CT even in cases that appear normal with conventional radiography[2,8,10] (Figs. 51D, 51E). However, some cervical pillar fractures may be difficult to detect with axial CT, especially when the fracture is in the horizontal plane.[5] These fractures may superficially resemble a normal or distracted facet joint.[11] CT demonstration of a distracted uncoverte-

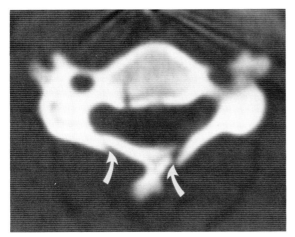

FIG. 51C. Unstable fracture of C4. Axial CT demonstrates fractures of both laminae of C4 (*arrows*) with some displacement causing spinal canal narrowing. A fracture of the vertebral body is also noted.

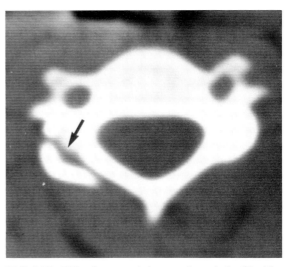

FIG. 51D. Pillar fracture. A fracture through the C3 pillar is present on the right (*arrow*).

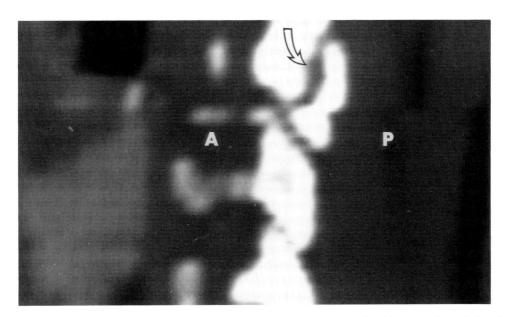

FIG. 51E. Pillar fracture. Same case as in Fig. 51D. Parasagittal reconstruction of axial images through the plane of the right articular processes. The fracture can be seen extending vertically through the C3 pillar (*arrow*). *A,* anterior; *P,* posterior.

bral joint or facet joint without other identified abnormalities suggests the need for further evaluation (e.g., reconstruction views, conventional tomography).[11] Uncovertebral joint subluxation indicates a rotation injury of the vertebral body and is usually accompanied by additional fracture or dislocation at that level. Facet joint subluxation or widening suggests the possibility of associated pillar fracture or locked facet[5,11] (Figs. 51F, 51G).

It is very important to study the conventional radiograph in conjunction with the CT study. Occasionally a unilateral locked facet is better demonstrated with conventional radiography than with CT.[8] The degree of vertebral body subluxation and compression can be better determined by conventional radiography. The CT evaluation of subluxation or compression depends on reconstruction views. Patient cooperation and detailed attention to technique

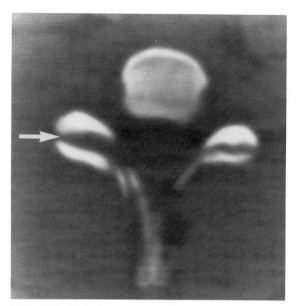

FIG. 51F. Widening of facet joint. There is considerable widening of the right facet joint (*arrow*). Compare with the normal left side. A fracture of the right lamina is present; however, this was better visualized on an adjacent scan.

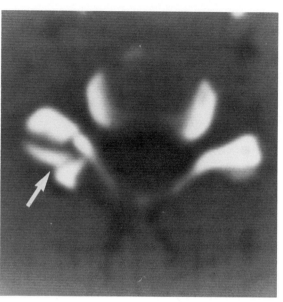

FIG. 51G. Pillar fracture. A comminuted pillar fracture of C6 (*arrow*) is present in this patient, who also had widening of the facet joint as shown in Fig. 51F. The CT demonstration of a pillar fracture may be misinterpreted as a widened facet joint in some cases.

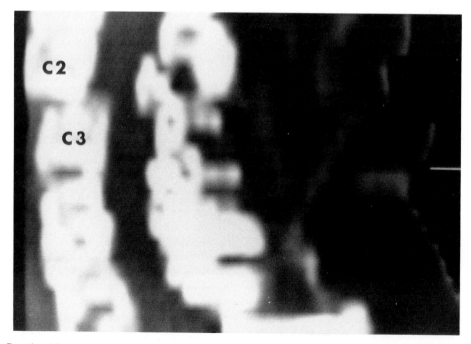

FIG. 51H. Pseudosubluxation. Sagittal reconstruction of axial images suggests a C2-C3 subluxation. However, no subluxation is present. The apparent subluxation on this reconstruction study occurred because the patient's position was altered during the examination by the placement of a pillow beneath the head and neck.

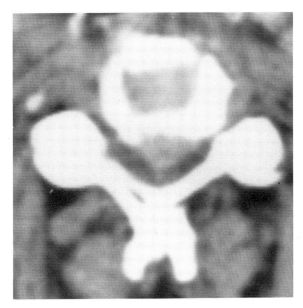

FIG. 51I. Posttraumatic disc herniation. This 71-year old was unable to move his extremities several hours after a fall on the stairs. Disc herniation at C3-C4 is present and is causing compression of the spinal cord.

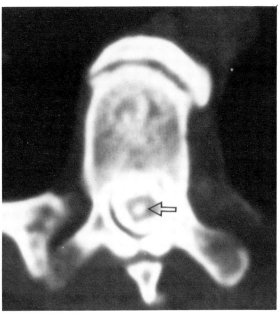

FIG. 51J. Posttraumatic syrinx. An intramedullary cavity (*arrow*) fills with intrathecal contrast in this patient, who had previously sustained a cervical fracture. The spinal cord is small and misshapen. (Figure courtesy of Jeffrey Blinder, M.D., Allentown, Pa.)

are necessary for satisfactory reconstruction. A sagittal reconstruction view is shown in Fig. 51H, which demonstrates an apparent subluxation of C2-C3. This patient had no history of trauma or arthritis and no actual cervical subluxation. This pseudosubluxation was created when a pillow was placed behind the patient's head and neck in the middle of the examination in an effort to relieve patient discomfort. With the patient's position altered during the examination, the sagittal reconstruction demonstrated an apparent subluxation.

The soft tissues of the cervical spine can best be visualized by CT. Acute traumatic disc herniation is not common but may occur in association with severe flexion or extension injury. It is important to recognize because it is a surgically correctable cause of cord compression (Fig. 51I). Surgical intervention may prevent a neurologic deficit if undertaken before spinal cord compromise or may reverse an already existing deficit.[7] Some investigators prefer CTM examination in the evaluation of acute spinal trauma unless there is a bullet injury or suspected epidural hematoma.[7,8] In addition to the previously discussed cervical nerve root avulsion and dural tear, the use of intrathecal water-soluble contrast improves the evaluation of spinal cord edema and disc herniation. In the evaluation of patients with late sequelae of spinal cord trauma, CTM is a valuable imaging modality for suspected syrinx[9] (Fig. 51J).

References

1. Brandt-Zawadzki M, Miller EM, Federle MP: CT in the evaluation of spine trauma. *AJR* 1981;136:369–375.
2. Coin CG, Pennink M, Ahmad WD, et al: Diving-type injury of the cervical spine: Contribution of computed tomography to management. *J Comput Assist Tomogr* 1979;3:362–372.
3. Miller MD, Gehweiler JA, Martinez S, et al: Significant new observations on cervical spine trauma. *AJR* 1978;130:659–663.
4. Morris RE, Hasso AN, Thompson JR, et al: Traumatic dural tears: CT diagnosis using metrizamide. *Radiology* 1984;152:443–446.
5. Pech P, Kilgore DP, Pojunas KW, et al: Cervical spinal fractures: CT detection. *Radiology* 1985;157:117–120.
6. Petras AF, Sobel DF, Mani JR, et al: CT myelography in cervical nerve root avulsion. *J Comput Assist Tomogr* 1985;9:275–279.
7. Post MJD, Green BA: The use of computed tomography in spinal trauma. *Radiol Clin North Am* 1983;21:327–375.
8. Post MJD, Green BA, Quencer RM, et al: The value of computed tomography in spinal trauma. *Spine* 1982;7:417–431.
9. Quencer RM, Green BA, Eismont FJ: Posttraumatic spinal cord cysts: Clinical features and characterization with metrizamide computed tomography. *Radiology* 1983;146:415–423.
10. Steppé R, Bellemans M, Boven F, et al: The value of computed tomography scanning in elusive fractures of the cervical spine. *Skeletal Radiol* 1981;6:175–178.
11. Yetkin Z, Osborn AG, Giles DS, et al: Uncovertebral and facet joint dislocations in cervical articular pillar fractures: CT evaluation. *AJNR* 1985;6:633–637.

CASE 52

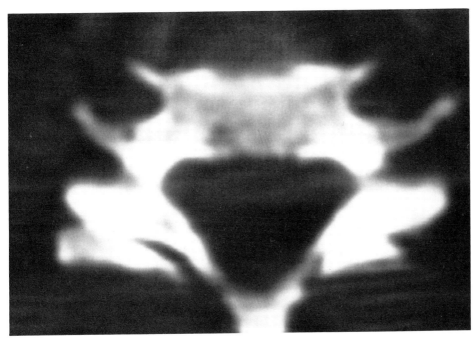

FIG. 52A. Axial CT at C6-C7. This 44-year-old trauma victim sustained injury to the cervical spine. Conventional radiographs demonstrated anterior subluxation of the C6 vertebral body in relation to C7.

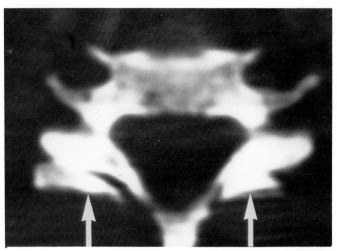

FIG. 52A-1. Bilateral locked facets. The superior articular processes of C7 (*arrows*) lie posterior to the inferior articular processes of C6. Note that the superior articular processes have a half-moon shape with a flat posterior margin. Normally, the inferior articular processes of the vertebra above lie posteriorly and have a rounded posterior configuration.

Bilateral Locked Facets

The superior articular processes of C7 lie posterior to the inferior articular processes of C6 (Fig. 52A-1). This is an example of bilateral locked facets. The injury is thought to occur from combined flexion and distraction forces, which cause separation of the spinous processes and disruption of the posterior ligaments.[3] The articular facets are separated and override. The inferior articular processes of the vertebra above become locked in position anterior to the superior articular processes of the vertebra below. Bilateral locking of the facets requires at least 50% anterior displacement of one vertebra upon the next lower vertebra.[1,4]

Bilateral locked facets have a typical CT appearance. Normally, the inferior articular processes of the vertebra above lie posterior to the superior articular processes of the vertebra below and have rounded posterior margins. When locking of the facets occurs, the half-moon-shaped superior articular processes lie posterior to the inferior articular processes and have flat posterior margins[6] (Fig. 52A-1). Parasagittal or oblique reconstruction through the plane of the articular processes further demonstrates the locked facet (Fig. 52B). Bilateral locked facets is considered an unstable injury and is frequently associated with spinal cord compromise.[1,4]

Unilateral locking of the facets occurs as a result of flexion, distraction, and rotation forces and is associated with less than 50% vertebral body displacement.[1] One facet joint acts as a fulcrum for the rotational forces. Simultaneous flexion and rotation lead to dislocation of the contralateral facet joint.[4]

The correct diagnosis of unilateral locked facet is overlooked on the initial radiographic examination in approximately 50% of cases.[2] On a true lateral radiograph, the left and right articular processes below the

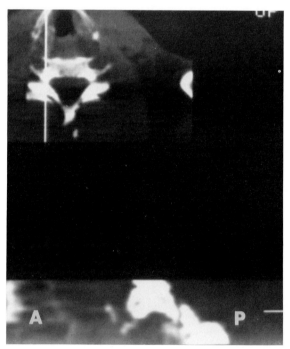

FIG. 52B. Locked facet. Same patient as in Fig. 52A-1. Parasagittal reconstruction through the plane of the right articular processes. Above: The plane of reconstruction is shown by the *white line*. Below: Reconstruction clearly demonstrates the anteriorly locked position of the inferior articular process of C6 in relation to the superior articular process of C7. *A*, anterior; *P*, posterior.

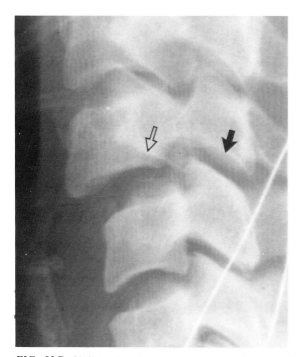

FIG. 52C. Unilateral locked facet. Lateral radiograph of the midcervical spine. There is a unilateral locked facet at C4-C5. One inferior articular process of C4 is dislocated anteriorly (*open arrow*) whereas the other remains in the normal position (*closed arrow*) articulating with a superior articular process of C5.

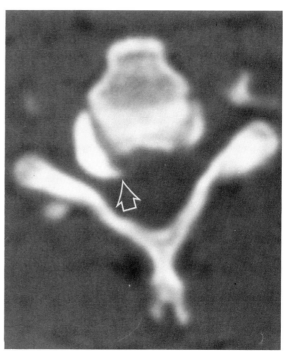

FIG. 52D. There is distraction of the uncovertebral joint on the right (*arrow*). This patient also had a unilateral locked facet. CT demonstration of uncovertebral joint distraction should raise the index of suspicion for a possible locked facet or pillar fracture.

dislocation lie symmetrically parallel to each other whereas above the dislocation two distinct sets of articular processes are seen at each level[5] (Fig. 52C). The processes above the level of dislocation appear to be in an oblique position because of the presence of rotation. Difficult diagnostic problems can be clarified by the CT study or conventional tomography. CT demonstrates normal articulation of the articular processes on one side and a reverse relationship of the processes on the opposite side, with the superior articular process seen posteriorly. Associated distraction of the uncovertebral joint (Fig. 52D) or fractures of the vertebral body and vertebral arch may be present[6] (Fig. 52E). The spinous process is rotated toward the abnormal side in one third of cases.[5] Unilateral locked facet is found in 16% of patients with trauma to the cervical spinal cord, with approximately 80% of these injuries occurring at C4-C5 or C5-C6.[5] It is considered a stable injury[1,2] and is less often associ-

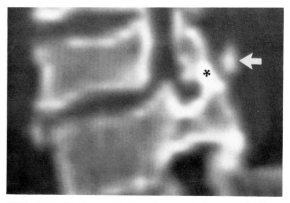

FIG. 52E. Unilateral locked facet and fracture. Oblique reconstruction through the plane of the articular processes demonstrates a locked facet. The superior articular process of C6 (*asterisk*) lies posterior to the inferior articular process of C5. An articular process fracture fragment is present posteriorly (*arrow*). Reconstruction views may be useful in evaluating complex pathology that is difficult to assess in the axial plane alone.

ated with neurologic deficit than is the bilateral locked facet injury.[3] However, if the diagnosis is not made promptly, reduction becomes difficult and recovery is hindered.[2,5]

In addition to demonstrating bilateral or unilateral locked facets, CT can be used to diagnose facet joint subluxation as well as perching of the facets. Facets are described as "perched" when the tip of the inferior articular process above rests on the tip of the superior articular process below. This injury may occur bilaterally or unilaterally. Axial CT examination reveals a "naked" facet with the superior and inferior articular processes imaged at different levels.[6]

References

1. Beatson TR: Fractures and dislocations of the cervical spine. *J Bone Joint Surg Br* 1963;45B:21–35.
2. Braakman R, Vinken PJ: Old luxations of the lower cervical spine. *J Bone Joint Surg Br* 1968;50B:52–60.
3. Gehweiler JA Jr, Osborne RL Jr, Becker RF: *The Radiology of Vertebral Trauma*. Philadelphia, WB Saunders, 1980.
4. Harris JH Jr; Acute injuries of the spine. *Semin Roentgenol* 1978;13:53–68.
5. Scher AT: Unilateral locked facet in cervical spine injuries. *AJR* 1977;129:45–48.
6. Yetkin Z, Osborn AG, Giles DS, et al: Uncovertebral and facet joint dislocations in cervical articular pillar fractures: CT evaluation. *AJNR* 1985;6:633–637.

CASE 53

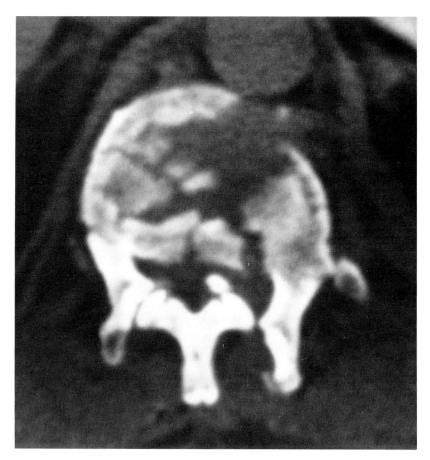

FIG. 53A. Axial CT at T12 in a 62-year-old male who developed paraplegia secondary to a fall from a height of 25 feet.

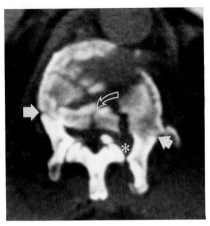

FIG. 53A-1. Burst fracture. There is a comminuted fracture of the vertebral body with posterior displacement of a fracture fragment into the spinal canal. Note the intrafragment fracture (*curved arrow*). Bilateral fracture of the pedicles (*straight arrows*) and dislocation of the left facet joint (*asterisk*) indicate the unstable nature of this injury. Spinal stenosis has been caused by fracture fragments derived from the vertebral body and from the posterior elements on the left.

Burst Fracture

This is an example of a burst fracture with comminuted fracture of the vertebral body, fracture of the pedicles, and dislocation of the left facet joint (Fig. 53A-1). Fracture fragments are displaced into the spinal canal, causing stenosis. In this case CT demonstrates the unstable nature of the fracture as well as the degree of spinal canal compromise.

The spine can be divided into three columns for purposes of evaluation of thoracolumbar fractures.[3,5,14] The anterior column consists of the anterior two thirds of the vertebral body, the anterior annulus, and the anterior longitudinal ligament. The middle column comprises the posterior third of the vertebral body, the posterior annulus, and the posterior longitudinal ligament. The posterior column includes the laminae, the spinous process, the articular processes, the facet joint capsules, the ligamenta flava, and the supraspinous and interspinous ligaments. Determination of the integrity of these three columns is useful in the evaluation of stability.

The simple anterior wedge fracture of the thoracolumbar spine involves the anterior column only, usually with less than 50% compression. This is a stable injury without neurologic deficit[14] and has a typical CT appearance with an arch of irregular bone density displaced circumferentially from the anterior portion of the vertebral body[12] (Fig. 53B). More than 50% compression of the anterior column may indicate additional injury to the posterior elements, with possible progression of spinal deformity and neurologic compromise.[5]

The more significant burst fracture is an axial compression injury with fracture of the vertebral endplate of sufficient force to cause the nucleus pulposus to be thrust into the vertebral body, producing comminution of the latter.[8] This leads to failure of the anterior and middle columns under compression.[3] The typical CT appearance consists of a single sagittally oriented fracture through the inferior endplate of the

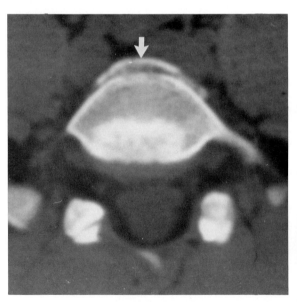

FIG. 53B. Simple anterior wedge fracture. A crescentic arc of bone (*arrow*) is displaced anteriorly from the vertebral body. No additional fracture is present. This is a characteristic CT appearance of a wedge fracture.

vertebral body (Fig. 53C) with broadening and comminution of the fracture as it is seen more superiorly[12,13] (Fig. 53D). Varying degrees of retropulsion of the posterior fracture fragments may be observed with compromise of the spinal canal (Fig. 53E). Typically, the fragment is derived from the superior posterior aspect of the vertebral body. The retropulsed fragment may have intrafragment fracture, anterior rotation, and/or cephalad or caudad displacement.[7] The fragment can be evaluated by CT, and the discovery of one or more of these findings may alert the surgeon to the possibility of increased difficulty in obtaining adequate reduction of the fragment.[7] The axial CT examination is supplemented by sagittal reconstruction views, which further delineate the fragment and the degree of spinal canal compromise (Figs. 53F, 53G). However, some authors believe that there is no correlation between the amount of preoperative spinal canal narrowing and the degree of neurologic impairment.[1,14,16] Coronal reconstruction graphically displays the bursting nature of this injury (Fig. 53H).

A fracture is unstable if it may lead to progressive increased spinal deformity or increased neurologic deficit. Although there is some controversy concerning the stability of different fractures, a burst fracture is clearly unstable when there is subluxation or dislocation of the articular processes or fracture-dislocation of the processes.[3,5] Some authors believe there is a high probability of progressive spinal deformity or neurologic compromise when the middle column has been disrupted and bone has been displaced into the canal.[3,5] Minimally displaced vertical fractures of the laminae, on the other hand, may not necessarily indicate instability.[9]

The initial evaluation of patients with suspected thoracolumbar injury begins with AP and lateral radiographs of the spine. The burst fracture is identified by conventional radiography; however, about 20% of posterior element fractures are not seen in this manner[11] and are better evaluated by CT.[13,14] In patients with significant x-ray or neurologic findings, the preliminary radiographs are followed by CT examination. CT is more useful than conventional radiography in helping to determine the presence of instability and determining the extent of spinal canal compromise.[1,11,15] CT evaluation for instability and spinal canal compromise adds important information, which, when used in conjunction with the clinical

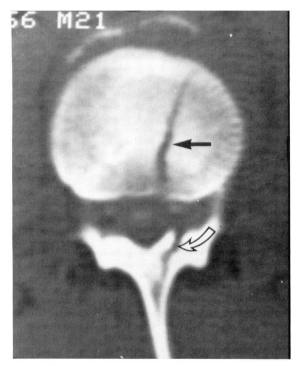

FIG. 53C. Burst fracture. CT scan at the inferior endplate of L3. The sagittally oriented fracture (*straight arrow*) at the inferior endplate is caused by propulsion of the nucleus pulposus into the vertebral body. In this scan section a fracture of the left lamina is also noted (*curved arrow*).

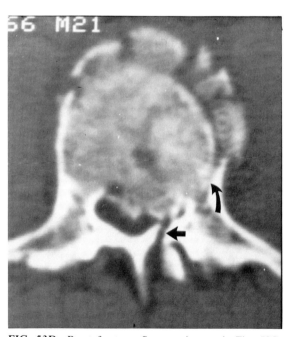

FIG. 53D. Burst fracture. Same patient as in Fig. 53C. This CT scan is at the level of the midbody of L3. Now the comminuted nature of the burst fracture is seen. There is a large fracture fragment displaced posteriorly into the spinal canal, causing severe compression of the thecal sac. Intrafragment fracture is present. Fractures of the left pedicle (*curved arrow*) and left lamina (*straight arrow*) can be seen.

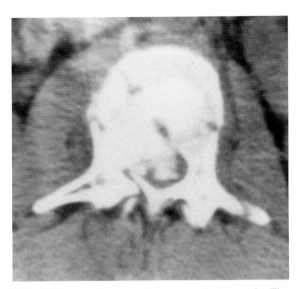

FIG. 53E. Burst fracture with severe spinal stenosis. The stenosis is caused by a large fracture fragment of the vertebral body displaced posteriorly into the canal. Intrafragment fracture is present. A displaced fracture of the right lamina is causing additional narrowing of the spinal canal. There is a fracture of the left transverse process.

assessment of the patient, helps determine the appropriate therapeutic approach. This includes the possible need for stabilization or decompression.

The use of CT in addition to conventional radiography eliminates the need for conventional tomography in the evaluation of most burst fractures.[2,4,11] This is advantageous because conventional tomography almost always requires repositioning of the patient into a lateral decubitus position, which may increase the neurologic deficit in a patient with an unstable fracture. The CT examination, on the other hand, is conducted with the patient in a supine position without need for excessive patient movement. CT is also more favorable than conventional tomography from the standpoint of radiation dosage received. The average radiation exposure for AP and lateral tomography is approximately 10 times that required for CT.[10] CTM may be reserved for patients suspected of having a dural rent or those with increasing symptoms not entirely explained by the osseous injury.[14]

CT can also be used in the follow-up examination of patients with burst fracture injury of the thoracolumbar spine. Patients who fail to recover from their partial neurologic deficit after Harrington distraction instrumentation can be examined by CT. The most common abnormality found is the presence of residual bone fragments within the canal.[6] However, some investigators have found no significant correlation between the degree of postsurgical improvement in spinal cross-sectional area and neurologic recovery.[16] Patients with complications occurring in the late recovery period may have CT demonstration of abundant callus compromising the canal or a lack of fusion of anterior vertebral body bone grafts.[6]

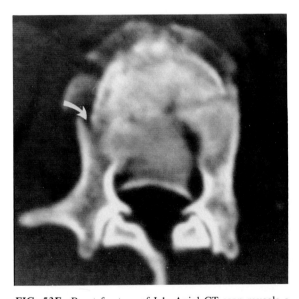

FIG. 53F. Burst fracture of L1. Axial CT scan reveals a comminuted vertebral body fracture with posterior displacement of a large fracture fragment. There is a fracture of the right pedicle (*arrow*). The left lamina appeared intact on adjacent scans, and the apparent defect seen in this figure is due to slight asymmetry in the scanning rather than fracture.

FIG. 53G. Burst fracture. Same case as in Fig. 53F. A sagittal reconstruction view after intrathecal introduction of water-soluble contrast demonstrates severe compression fracture of L1 with posterior displacement of fragment (*arrow*) derived from the superior aspect of the vertebral body. There is compression of the subarachnoid space with partial block of the contrast at the fracture site. The anterior (*A*) and posterior (*P*) relationships are marked.

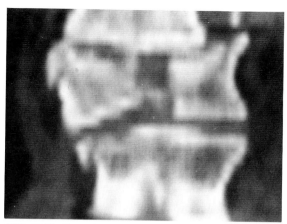

FIG. 53H. Burst fracture. Coronal reconstruction demonstrates the bursting quality of this injury. There is midline cleavage of the vertebral body with lateral displacement of the vertebral body halves.

References

1. Brant-Zawadzki M, Jeffrey RB, Jr, Minagi H, et al: High resolution CT of thoracolumbar fractures. *AJNR* 1982;3:69–74, *AJR* 1982;138:699–704.
2. Brant-Zawadzki M, Miller EM, Federle MP: CT in the evaluation of spine trauma. *AJR* 1981;136:369–375.
3. Denis F: Spinal instability as defined by the three-column spine concept in acute spinal trauma. *Clin Orthop* 1984;189:65–76.
4. Durward QJ, Schweigel JF, Harrison P: Management of fractures of the thoracolumbar and lumbar spine. *Neurosurgery* 1981;8:555–560.
5. Ferguson RL, Allen BL Jr: A mechanistic classification of thoracolumbar spine fractures. *Clin Orthop* 1984;189:77–88.
6. Golimbu C, Firooznia H, Rafii M, et al: Computed tomography of thoracic and lumbar spine fractures that have been treated with Harrington instrumentation. *Radiology* 1984;151:731–733.
7. Guerra J Jr, Garfin SR, Resnick D: Vertebral burst fractures: CT analysis of the retropulsed fragment. *Radiology* 1984;153:769–772.
8. Holdsworth F: Fractures, dislocations, and fracture-dislocations of the spine. *J Bone Joint Surg Am* 1970;52-A:1534–1551.
9. Jacobs RR, Casey MP: Surgical management of thoracolumbar spinal injuries: General principles and controversial considerations. *Clin Orthop* 1984;189:22–35.
10. Keene JS: Radiographic evaluation of thoracolumbar fractures. *Clin Orthop* 1984;189:58–64.
11. Keene JS, Goletz TH, Lilleas F, et al: Diagnosis of vertebral fractures: A comparison of conventional radiography, conventional tomography, and computed axial tomography. *J Bone Joint Surg Am* 1982;64-A:586–595.
12. Kilcoyne RF, Mack LA, King HA, et al: Thoracolumbar spine injuries associated with vertical plunges: Reappraisal with computed tomography. *Radiology* 1983;146:137–140.
13. Lindahl S, Willen J, Nordwall A, et al: The crush-cleavage fracture: A new thoracolumbar unstable fracture. *Spine* 1983;8:559–569.
14. McAfee PC, Yuan HA, Fredrickson BE, et al: The value of computed tomography in thoracolumbar fractures: An analysis of one hundred consecutive cases and a new classification. *J Bone Joint Surg Am* 1983;65-A:461–473.
15. Post MJD, Green BA, Quencer RM, et al: The value of computed tomography in spinal trauma. *Spine* 1982;7:417–431.
16. Shuman WP, Rogers JV, Sickler ME, et al: Thoracolumbar burst fractures: CT dimensions of the spinal canal relative to postsurgical improvement. *AJR* 1985;145:337–341.

CASE 54

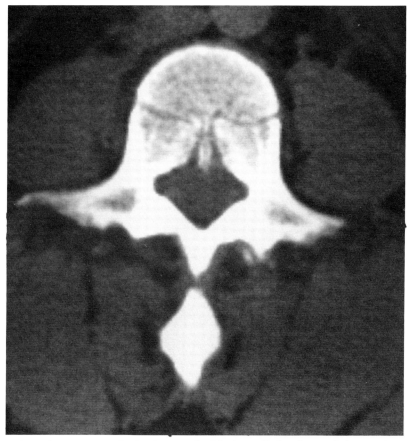

FIG. 54A Axial CT of a lumbar vertebral body studied at bone window settings. This patient sustained trauma to the back.

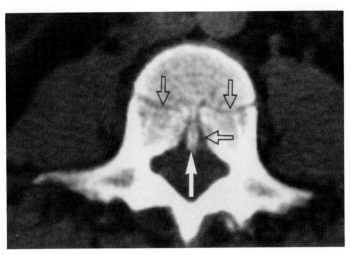

FIG. 54A-1. Groove of the basivertebral vein. Axial CT demonstrates a Y-shaped radiolucency (*short arrows*) that represents a vascular groove for the basivertebral vein. This should not be mistaken for a fracture. A bony septum (*long arrow*) separates the paired posterior venous channels.

Basivertebral Veins

The Y-shaped lucency demonstrated on axial CT is caused by vertebral channels containing the basivertebral vein (Fig. 54A-1). This typical appearance of the groove of the basivertebral vein may simulate a fracture. Other venous structures may appear similar to a disc herniation, whereas still others may simulate an osteophyte.

A complex system of freely communicating intraspinal and extraspinal veins is present throughout all spinal levels[1] (Fig. 54B). A portion of the vertebral venous drainage courses posteriorly at the midvertebral body level through the basivertebral vein and empties into the retrovertebral plexus of veins (RPV)[2–5] (Figs. 54B–54D). The RPV lies between the vertebral body and the posterior longitudinal ligament and is joined to similar plexuses located cephalad and caudad by longitudinally oriented anterior internal vertebral veins (AIVV). Radicular (pedicular, intervertebral) veins arise from the AIVV and pass through the neural foramen to join the ascending lumbar vein.

The various venous structures can be identified with high-resolution CT. The basivertebral vein courses horizontally at the midvertebral level along Y- or V-shaped channels that appear radiolucent with axial CT (Fig. 54A-1). Besides the typical configuration and midvertebral body location, there are other features of basivertebral channels that differentiate them from fracture. The basivertebral groove may have a distinct osseous wall that is not found in the CT evaluation of a fracture.[7] In addition, fractures are typically demonstrated on contiguous scans, whereas the basivertebral vein channels are usually not apparent on adjacent scans except for the prominent posterior portion.[7] Posteriorly, there may be single or paired channels (Fig. 54E) as well as an interposed osseous septum

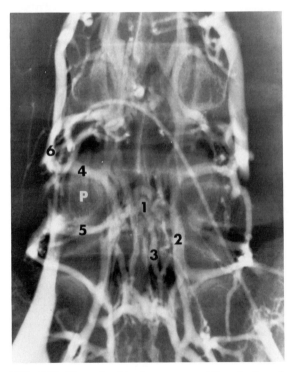

FIG. 54B. Normal epidural venogram, AP projection. *1*, retrovertebral plexus of veins; *2*, lateral anterior internal vertebral vein; *3*, medial anterior internal vertebral vein; *4*, superior radicular (pedicular, intervertebral) vein; *5*, inferior radicular (pedicular, intervertebral) vein; *6*, ascending lumbar vein; *P*, pedicle. (Figure courtesy of Dr. Raziel Gershater, Willowdale, Ontario, Canada.)

that sometimes projects slightly into the spinal canal. The bony septum is a normal structure seen at the midvertebral level and should not be mistaken for an osteophyte, which typically occurs at the vertebral margins (Fig. 54F).

The AIVV are readily identified on CT in the lower lumbar spine because of an abundance of contrasting epidural fat that surrounds the veins. These veins may appear as several small, round soft-tissue densities located bilaterally in the anterior epidural space (Fig. 54G). Knowledge of their appearance should prevent the mistaken diagnosis of disc herniation.[6] These epidural veins are not readily identified in the cervical and thoracic region because of a paucity of epidural fat at these levels. Radicular veins are occasionally identified on CT and can be differentiated from nerve roots by their thinner width and more horizontal course[5] (Fig. 54H).

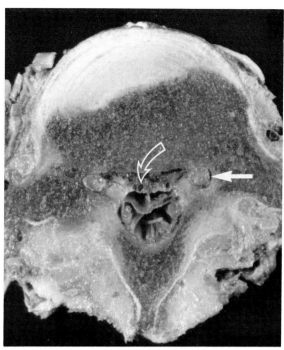

FIG. 54D. Gross anatomic specimen of a lumbar vertebra in a near-axial plane. This section is at the level of the lateral recess. The anterior internal vertebral veins (*curved arrow*) are surrounded by fat within the epidural space. The nerve root can be identified (*straight arrow*) within the lateral recess.

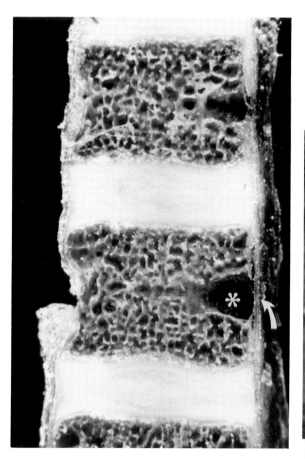

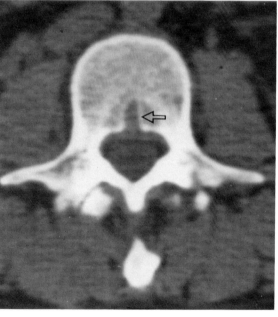

FIG. 54C. Gross anatomic specimen of the lumbar vertebrae in the sagittal plane. The basivertebral vein is prominent posteriorly (*asterisk*). Note the relationship of the basivertebral vein and the posterior longitudinal ligament (*arrow*).

FIG. 54E. Normal posterior groove for the basivertebral vein. Axial CT through the midportion of a lumbar vertebral body. A prominent single basivertebral groove is noted posteriorly (*arrow*). This normal appearance should not be confused with an osteolytic lesion.

Case 54

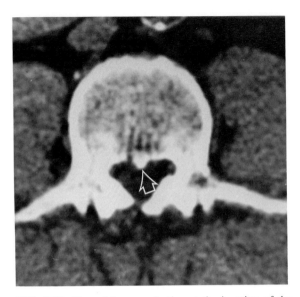

FIG. 54F. Normal bony projection at the junction of the basivertebral vein and retrovertebral plexus of veins. Axial CT through the midportion of a lumbar vertebral body. The bony septum appears as a bony prominence (*arrow*) at the midvertebral body level. This should not be confused with an osteophyte, which typically occurs at the level of the superior and inferior vertebral margins.

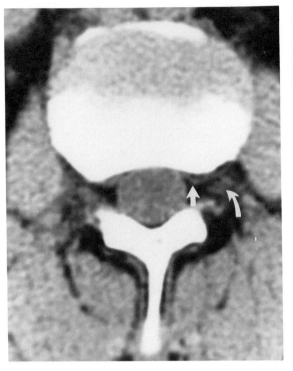

FIG. 54H. Normal radicular vein. Axial CT through a lumbar vertebra at the level of the neural foramen. The radicular vein (*straight arrow*) is thinner and has a more horizontal course than the nerve root (*curved arrow*).

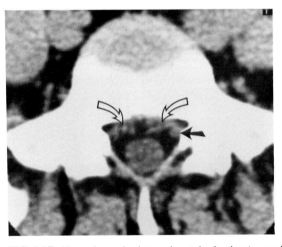

FIG. 54G. Normal anterior internal vertebral veins (*curved arrows*) appear in cross section as small, rounded densities within the anterior epidural space. The veins are well visualized because they are surrounded by fat. The larger S1 nerve root (*straight arrow*) is readily differentiated from the venous structures.

References

1. Batson OV: The vertebral vein system. *AJR* 1957;78:195–212.
2. Crock HV, Goldwasser M: Anatomic studies of the circulation in the region of the vertebral end-plate in adult greyhound dogs. *Spine* 1984;9:702–706.
3. Crock HV, Yoshizawa H: The blood supply of the lumbar vertebral column. *Clin Orthop* 1976;115:6–21.
4. Crock HV, Yoshizawa H, Kame SK: Observations on the venous drainage of the human vertebral body. *J Bone Joint Surg Br* 1973;55-B:528–533.
5. Haughton VM, Syvertsen A, Williams AL: Soft-tissue anatomy within the spinal canal as seen on computed tomography. *Radiology* 1980;134:649–655.
6. Meijenhorst GCH: Computed tomography of the lumbar epidural veins. *Radiology* 1982;145:687–691.
7. Sartoris DJ, Resnick D, Guerra J Jr: Vertebral venous channels: CT appearance and differential considerations. *Radiology* 1985;155:745–749.

CASE 55

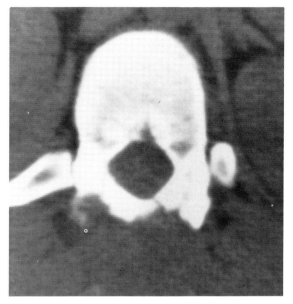

FIG. 55A. Axial CT of T12 at the level of the pedicles. This patient sustained a back injury during a motor vehicle accident. Conventional radiographs demonstrated wide separation of the spinous processes of T11 and T12.

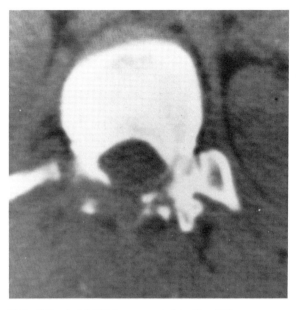

FIG. 55B. Axial CT 4 mm caudad to Fig. 55A.

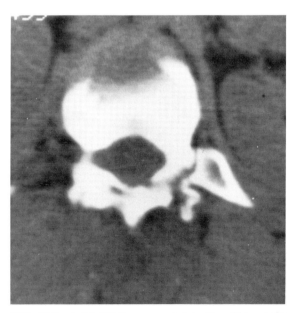

FIG. 55C. Axial CT 8 mm caudad to Fig. 55A, at the inferior aspect of T12.

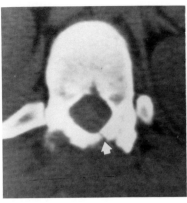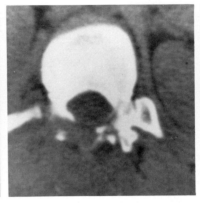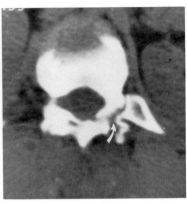

FIG. 55A-1. Horizontal fracture with diastasis (Smith fracture). There is a fracture of the left lamina of T12 (*arrow*).

FIG. 55B-1. Four millimeters caudad to Fig. 55A-1 the laminae are no longer visualized as solid bony structures. Fragments of bone can be seen.

FIG. 55C-1. Eight millimeters caudad to Fig. 55A-1 the laminae are once again demonstrated. A fracture of the left lamina is present (*arrow*). The sequential evaluation of these three CT scans suggests a horizontal fracture through the posterior elements with diastasis. Portions of the laminae are displaced both cephalad (Fig. 55A-1) and caudad (Fig. 55C-1) to the site of horizontal fracture (Fig. 55B-1). Conventional radiographs and sagittal reconstruction views had demonstrated wide separation of the spinous processes of T11 and T12, indicating disruption of the posterior ligamentous complex.

Horizontal Fracture With Diastasis

The transaxial scans reveal a fracture of the left lamina of T12. As the scans are studied sequentially at 4-mm intervals, it becomes evident that the laminae seem to disappear in the middle section although they are clearly seen just above and below (Figs. 55A-1–55C-1). The "disappearing laminae" sign is due to horizontal fracture through the laminae, with diastasis occurring parallel to the plane of the transaxial CT scan. The fracture is further clarified by oblique reconstruction views through the lamina (Fig. 55D). Additional parasagittal reconstruction of the axial images through the plane of the pedicle reveals a horizontal pedicle fracture (Fig. 55E). Widening of the space between the spinous processes of T11 and T12 was seen on conventional radiographs and midsagittal CT reconstruction views and indicates disruption of the posterior ligamentous complex.

The usual flexion-compression injury of the thoracolumbar spine results in forces being directed through the anterior portion of the vertebral body. This leads to anterior compression wedging of the vertebral body. A more unusual flexion injury is the Chance fracture, a horizontal fracture through the spinous process and neural arch with additional fracture of the posterior superior aspect of the vertebral body[1] (Fig. 55F). Years after this injury was first described, a similar fracture was noted in association with a lap-type seat belt injury.[3] This occurs when a victim of a motor vehicle accident wearing a lap-type seat belt experiences sudden deceleration. The victim's body is forced into flexion with the seat belt and the abdominal wall forming the fulcrum point at impact.[3,5] This leads to tension forces being directed through the posterior structures such as seen in a Chance fracture. However, the forces may cause ligamentous disruption or combined bony and ligamentous injury. The combined injury (such as shown in the lead case) has been termed a Smith fracture and consists of horizontal fracture through the pedicles, laminae, and transverse processes along with disruption of the posterior ligamentous complex.[2,5] This type of injury may also have fractures of the superior articular processes and avulsion of the superior or inferior portion of the posterior vertebral body[2,5] (Fig. 55G).

The diagnosis of a seat belt fracture can be made with conventional radiography. Both the AP and lateral radiographs are studied for clues to this injury. On the AP radiograph a break may be seen in the continuity of the cortex of the pedicles and/or the spinous processes.[4] Also on the AP view, distraction forces lead to separation and elevation of the posterior elements, with posterior elements no longer superimposed over the corresponding vertebral body.[4] On the lateral radiograph the horizontal fracture through the posterior structures may be identified. The spinous processes are widely separated if ligamentous

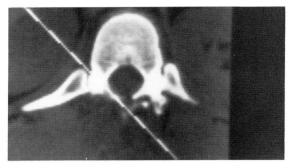

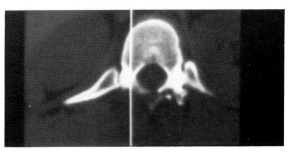

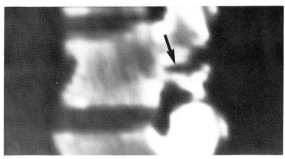

FIG. 55D. Horizontal fracture with diastasis (Smith fracture). Same case as in Fig. 55A-1. Oblique reconstruction through the plane of the right lamina. Above: Axial scan with cursor line showing plane of reconstruction. Below: Oblique reconstruction demonstrates horizontal fracture of the lamina with diastasis (*arrow*).

FIG. 55E. Horizontal fracture wth diastasis (Smith fracture). Same case as in Fig. 55A-1. Parasagittal reconstruction through the plane of the right pedicle. Above: Axial scan with cursor line showing plane of reconstruction. Below: Parasagittal reconstruction demonstrates horizontal fracture of the pedicle (*arrow*).

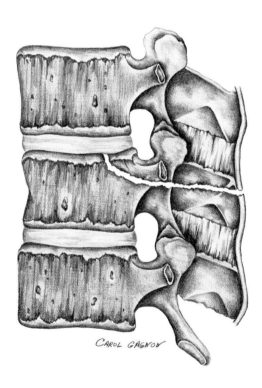

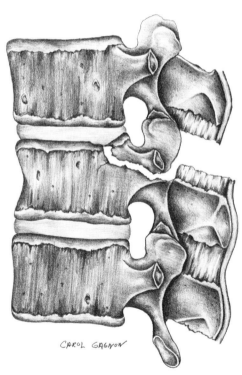

FIG. 55F. Chance fracture. Drawing depicts horizontal fracture of the lamina and pedicle as well as fracture of the vertebral body. The interspinous distance is normal because of intact posterior ligaments.

FIG. 55G. Smith fracture. Drawing demonstrates horizontal fracture of the lamina and pedicle as well as fracture of the vertebral body. In this injury there is widening of the distance between spinous processes. This occurs as a result of disruption of the posterior ligamentous complex.

injury is present. An optimal lateral view, however, is often difficult to obtain in the injured patient, and a CT scan may be performed to better delineate the abnormalities. The finding of disappearing laminae on the transaxial scan is a clue to the presence of a horizontal fracture through the laminae with associated diastasis. This finding suggests that additional reconstruction views should be obtained with particular attention provided to the posterior elements. For this purpose parasagittal reconstruction through each pedicle and oblique reconstruction through each lamina are recommended. Both the Chance fracture and the Smith fracture demonstrate horizontal fractures through the pedicles and laminae. The spinous process is fractured horizontally in the Chance fracture but remains intact in the Smith fracture, in which the posterior force is directed through the ligaments. The ligamentous disruption that occurs with a Smith fracture leads to wide separation of the neighboring spinous processes and results in a more unstable injury than the Chance fracture.

References

1. Chance GQ: Note on a type of flexion fracture of the spine. *Br J Radiol* 1948;21:452–453.
2. Gehweiler JA Jr, Osborne RL, Becker RF: *The Radiology of Vertebral Trauma*. Philadelphia, WB Saunders, 1980.
3. Howland WJ, Curry JL, Buffington CB: Fulcrum fractures of the lumbar spine: Transverse fracture induced by an improperly placed seat belt. *JAMA* 1965;193:240–241.
4. Rogers LF: The roentgenographic appearance of transverse or Chance fractures of the spine: The seat belt fracture. *AJR* 1971;111:844–849.
5. Smith WS, Kaufer H: Patterns and mechanisms of lumbar injuries associated with lap seat belts. *J Bone Joint Surg Am* 1969;51-A:239–254.

CASE 56

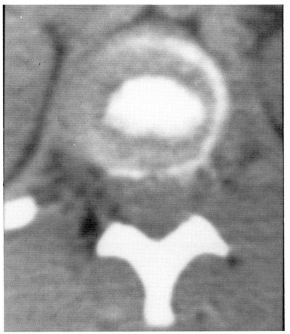

FIG. 56A. Axial CT scan at the level of the T12-L1 intervertebral disc. This 14-year-old patient was injured in a motor vehicle accident.

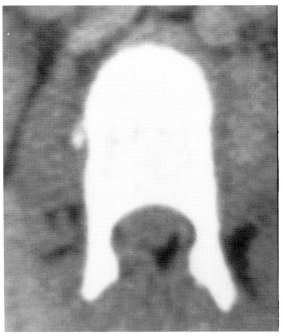

Fig. 56B. Axial CT scan obtained 12 mm caudad to Fig. 56A.

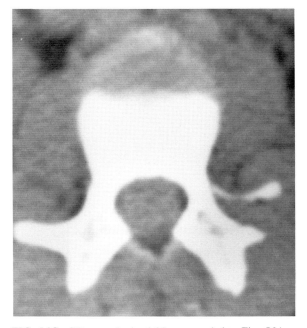

FIG. 56C. CT scan obtained 20 mm caudad to Fig. 56A.

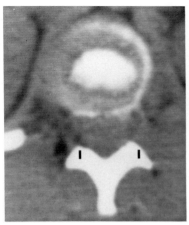 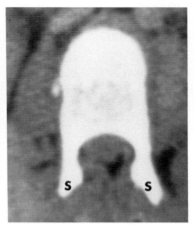 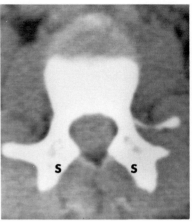

FIG. 56A-1. Facet distraction injury. The inferior articular processes of T12 (*I*) are seen without the companion superior articular processes of L1.

FIG. 56B-1. The superior articular processes of L1 (*S*) are seen lying "naked" without their companion inferior articular processes of T12.

FIG. 56C-1. The naked facet is still evident 2 cm below Fig. 56A-1. There is a 2-cm separation of the superior articular processes (*S*) from the inferior articular processes. This separation is due to disruption of the posterior ligamentous complex.

Facet Distraction Injury: Naked Facet

The axial CT scans demonstrate the "naked facet" sign of vertical distraction of the articular processes caused by disruption of the posterior ligamentous complex.[3] The inferior articular processes of the T12 vertebral body lie "naked" (Fig. 56A-1) without their companion superior articular processes of L1, which are seen on the scans obtained 1 and 2 cm caudad (Figs. 56B-1, 56C-1). Sagittal reconstruction can be used to further demonstrate wide separation of the spinous processes and to evaluate additional vertebral body compression and subluxation (Fig. 56D). This injury, like the Chance and Smith fractures previously described, occurs secondary to forced flexion about a fulcrum point centered at the abdominal wall such as occurs with a lap-type seat belt injury.[5] However, in the facet distraction injury the pedicles, laminae, and spinous process remain intact. The force is directed through the posterior ligamentous complex, resulting in vertical distraction of the articular processes[3,5] (Fig. 56E). This represents an acutely unstable injury requiring reduction and operative fixation to prevent late instability.[3]

In patients with unstable injuries of the thoracolumbar spine, conventional radiography detects widening of the interspinous space in 50% of cases and widening of the apophyseal joints in 30%.[2] These percentages are low because of the difficulty in visualizing the posterior structures by conventional radiography.[2] CT more clearly demonstrates the abnormalities of ligamentous injury and the presence or absence of associated posterior element fracture. The transaxial naked facet sign and the sagittal reconstruction demonstration of widening of the interspinous space both indicate disruption of the posterior ligamentous complex (Figs. 56F, 56G). In addition, CT in the axial, sagittal, and coronal planes demonstrates

FIG. 56D. Facet distraction injury. Sagittal reconstruction. Same patient as in Fig. 56A-1. There is compression and subluxation of the L1 vertebral body. Note the wide separation of the spinous processes of T12 and L1.

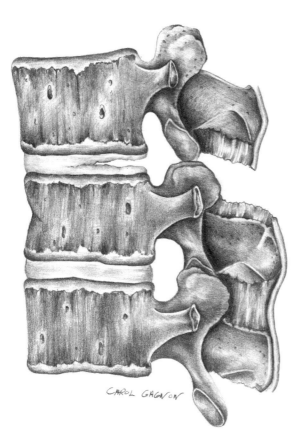

FIG. 56E. Facet distraction injury. Vertical distraction of the articular and spinous processes is due to posterior ligamentous injury. Vertebral body compression may be mild, as in this drawing. The horizontal force is directed through the disc and posterior ligaments without causing fracture of the pedicles, laminae, or spinous process.

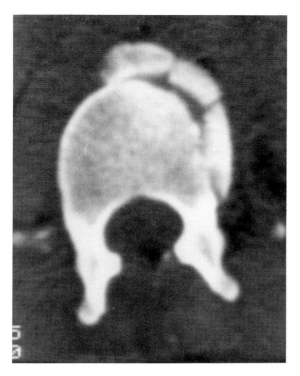

FIG. 56F. Facet distraction injury. Axial CT through the L1 vertebral body reveals the naked facet sign with superior articular processes of L1 lying free of their companion inferior articular processes of T12. Fracture of the L1 vertebral body is noted. This 23-year-old was in a motor vehicle accident.

FIG. 56G. Facet distraction injury. Same patient as in Fig. 56F. This sagittal reconstruction of the axial images demonstrates wide separation of the spinous processes of T12 and L1. There is compression of the L1 vertebral body and subluxation at T12-L1. This unstable injury was surgically reduced and stabilized with rods and bony fusion.

the presence of vertebral body compression (which is frequently mild), bone encroachment on the spinal canal from a vertebral body fracture fragment, subluxation, and kyphosis.[3]

When considered as a group, Chance fracture, Smith fracture, and facet distraction injury may be considered lap-type seat belt injuries, although they may occur in other clinical settings in which the abdominal wall encounters a fixed object while the upper body is forced into flexion (e.g., a fall or jump onto an object such as a fence). These spinal injuries have associated abdominal trauma in 15% of cases.[4] The abdominal injuries include tear of the small bowel and/or mesentery, perforation of the duodenum, or laceration of a solid viscus.[1,4] The spinal and abdominal injuries are not found when a shoulder harness seat belt is used (although other injuries related to the upper thorax may occur). This syndrome, although associated with the use of seatbelts, should not in any way discourage the use of these restraining devices; however, the physician caring for victims of motor vehicle accidents must be aware of these injuries to ensure proper diagnosis and therapy.

References

1. Dehner JR: Seatbelt injuries of the spine and abdomen. *AJR* 1971;111:833–843.
2. Gehweiler JA Jr, Daffner RH, Osborne RL Jr: Relevant signs of stable and unstable thoracolumbar vertebral column trauma. *Skeletal Radiol* 1981;7:179–183.
3. O'Callaghan JP, Ullrich CG, Yuan HA, et al: CT of facet distraction in flexion injuries of the thoracolumbar spine: The naked facet. *AJR* 1980;134:563–568.
4. Rogers LF: The roentgenographic appearance of transverse or Chance fractures of the spine: The seat belt fracture. *AJR* 1971;111:844–849.
5. Smith WS, Kaufer H: Patterns and mechanisms of lumbar injuries associated with lap seat belts. *J Bone Joint Surg Am* 1969;51-A:239–254.

CASE 57

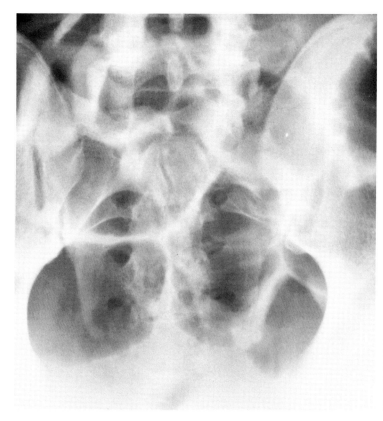

FIG. 57A. AP view of the sacrum extracted from an AP radiograph of the entire pelvis. This was obtained at the time of initial evaluation of a patient with multiple trauma, which included fractures of the pubic bones.

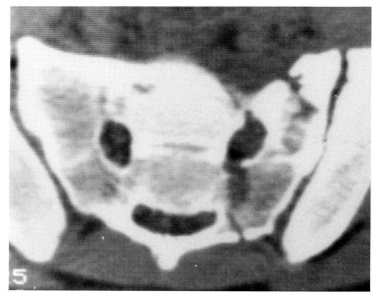

FIG. 57B. CT of the sacrum obtained after the patient's condition had stabilized.

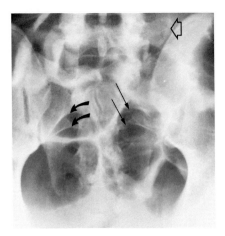
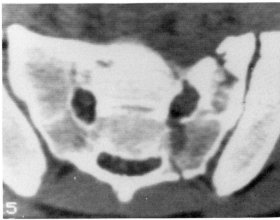

FIG.57A-1. Sacral fracture. The arcuate lines of the sacrum have a normal smooth, gently curved arc on the right (*curved arrows*). Fracture on the left is indicated by the lack of regular continuity of the arcuate lines (*closed straight arrows*). This area is difficult to evaluate because of overlying bowel. Note also the fracture of the transverse process of L5 on the left (*open arrow*).

FIG. 57B-1. CT demonstration of the sacral fracture is dramatic. The fracture extends from the anterior cortex to the posterior cortical surface. There is fracture into the left sacral foramen.

Sacral Fracture

Fractures of the sacrum may be difficult to visualize with conventional radiography because of the curved orientation of the sacrum and the presence of overlying soft-tissue structures and bowel gas (Fig. 57A-1). Additional angled and oblique radiographs and conventional tomography may be used to further evaluate the sacrum. However, CT has been found to be a useful modality to detect sacral fractures, determine their extent, and identify additional soft-tissue abnormalities such as pelvic hematoma[5,12] (Figs. 57B-1, 57C). The CT study can be done in the supine position and does not require patient movement.

Sacral fractures occur in association with fractures of the anterior arch of the pelvis in 90% of cases, often in patients suffering additional injuries.[11] These sacral fractures usually have an oblique or a vertical orientation. The mechanism of injury is either AP or lateral compression of the pelvic ring or vertical shear.[11] Severe trauma to one leg or one side of the body leads to fracture of the ischiopubic rami or diastasis of the symphysis pubis anteriorly. The force of injury is transmitted posteriorly, causing fracture of the sacrum or ilium or diastasis of the sacroiliac joint.[10,11] Thus, whenever anterior arch fractures of the pelvis are present, careful attention must be given to possible posterior arch injury. Fracture of a transverse process of L5 may act as an additional clue to an underlying fracture of the sacrum or diastasis of the sacroiliac joint.[9] The clinical significance in identifying posterior arch fractures is that fractures that involve both the anterior and the posterior arch are more unstable and are associated with increased morbidity compared with fractures limited to the anterior arch.[10] Only 10% of sacral fractures occur without additional pelvic fractures.[11] These isolated sacral fractures are usually transverse and are frequently as-

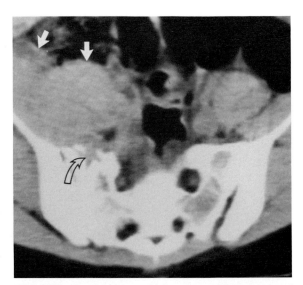

FIG. 57C. Sacral fracture with hematoma. There is a sacral fracture (curved arrow) associated with a large hematoma (straight arrows). This patient had an additional fracture of the right ilium, which was seen on adjacent scans. CT is an excellent modality for evaluation of soft-tissue pathology such as hematoma.

sociated with neural injury, which may lead to rectal and bladder incontinence.[1,11,12]

The critically injured multiple-trauma victim is routinely examined with an AP radiograph of the pelvis as part of the initial evaluation following chest and cervical spine radiography.[1] Fractures of the sacrum are often subtle, with buckling of a superior foraminal line (arcuate line) the only clue to the injury.[7] More than 60% of sacral fractures may be overlooked on the initial radiographic examination of patients with multiple fractures of the pelvic ring.[7] In one report, CT examinations of the pelvis were performed for patients with multiple injuries including pelvic fractures.[5] The studies were done within the first 4 days of injury at a time when the patient was clinically stable. In 20% of cases the CT scan more clearly delineated the size, configuration, and displacement of radiographically apparent fractures. In an additional 65% of cases the CT study was considered "extremely helpful."[5] In these cases additional fractures of the sacrum or diastasis of the sacroiliac joint were found although they had not been identified on the pelvic radiographs. Other authors believe, however, that although CT is valuable in evaluating the deformity of the sacral canals and posterior elements, it is not essential in diagnosing sacral fracture if the conventional radiograph is carefully studied for possible disruption, displacement, deformity, and density changes of the sacral arcuate lines.[7]

Occasionally a sacral fracture occurs either spontaneously or after minimal trauma in an elderly patient with osteopenia or in a patient who has received radiation therapy to the pelvis (Fig. 57D). In this situation the radionuclide bone scan may be positive and conventional radiographs negative. CT is then useful in diagnosing a fracture and excluding other disorders such as metastasis, which might be a diagnostic consideration in an elderly patient presenting with pain and a positive bone scan.[6] The CT scan demonstrates a vertically oriented fracture line with marginal reactive sclerosis.[3] The fracture is typically in the sacral alae, parallel to the sacroiliac joint, just lateral to the margins of the lumbar spine.[2] Insufficiency fractures of the sacrum may occur in association with insufficiency fractures of the pubic bone or ileum.[2,3]

CT examination of the pelvis may detect diastasis of the sacroiliac joint that is sometimes not appreciated on the AP radiograph of the pelvis. The width of the sacroiliac joint can be measured by CT and is normally 2.5 to 4 mm.[8] A CT study of posttraumatic sacroiliac joint widening is shown in a patient with multiple fractures of the anterior pelvic arch (Fig. 57E). In children with subchondral fractures of the iliac bone, the conventional radiographs and CT scans obtained immediately after trauma may have an appearance suggesting sacroiliac joint widening.[4] Within several days CT scanning reveals the true nature of the injury with demonstration of a normal joint and subchondral new bone at the fracture site.[4]

References

1. Ayella RJ: *Radiologic Management of the Massively Traumatized Patient.* Baltimore, Williams & Wilkins, 1978.
2. Cooper KL, Beabout JW, Swee RG: Insufficiency fractures of the sacrum. *Radiology* 1985;156:15–20.

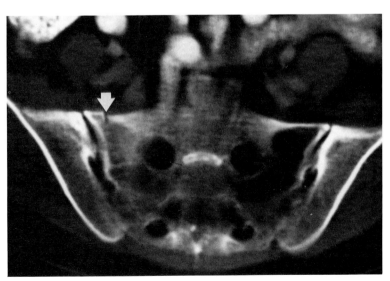

FIG. 57D. Sacral fracture. This elderly female with osteopenia sustained a fracture of the sacrum after minor trauma. CT demonstrates the subtle fracture on the anterior surface of the sacrum (*arrow*).

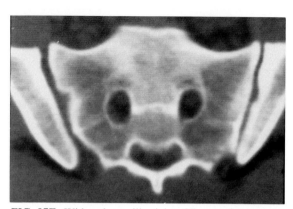

FIG. 57E. Widened sacroiliac joint. CT scan of the sacrum demonstrates widening of the right sacroiliac joint after trauma. Compare with the normal left side. The width of the sacroiliac joints is normally 2.5 to 4.0 mm.

3. DeSmet AA, Neff JR: Pubic and sacral insufficiency fractures: Clinical course and radiologic findings. *AJR* 1985;145:601–606.
4. Donoghue V, Daneman A, Krajbich I, et al: CT appearance of sacroiliac joint trauma in children. *J Comput Assist Tomogr* 1985;9:352–356.
5. Dunn EL, Berry PH, Connally JD: Computed tomography of the pelvis in patients with multiple injuries. *J Trauma* 1983;23:378–382.
6. Gacetta DJ, Yandow DR: Computed tomography of spontaneous osteoporotic sacral fractures: Case report. *J Comput Assist Tomogr* 1984;8:1190–1191.
7. Jackson H, Kam J, Harris JH Jr, et al: The sacral arcuate lines in upper sacral fractures. *Radiology* 1982;145:35–39.
8. Lawson TL, Foley WD, Carrera GF, et al: The sacroiliac joints: Anatomic, plain roentgenographic, and computed tomographic analysis. *J Comput Assist Tomogr* 1982;6:307–314.
9. Rogers LF: Common oversights in the evaluation of the patient with multiple injuries. *Skeletal Radiol* 1984;12:103–111.
10. Rogers LF: *Radiology of Skeletal Trauma.* New York, Churchill Livingstone, 1982.
11. Schmidek HH, Smith DA, Kristiansen TK: Sacral fractures. *Neurosurgery* 1984;15:735–746.
12. Weaver EN Jr, England GD, Richardson DE: Sacral fracture: Case presentation and review. *Neurosurgery* 1981;9:725–728.

CASE 58

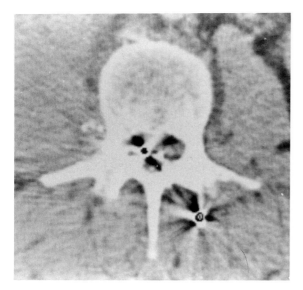

FIG. 58A. This patient sustained a gunshot injury. Axial CT at L3.

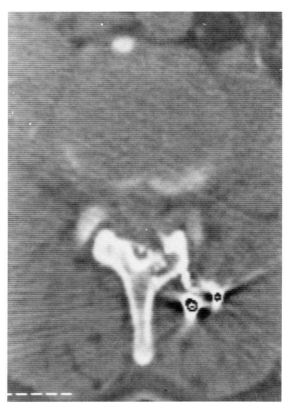

FIG. 58B. Axial CT at L3-L4.

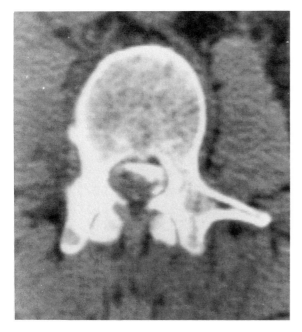

FIG. 58C. Axial CTM obtained 9 days later when the patient's neurologic condition had worsened. CTM was performed after a block was demonstrated with myelography. What information is gained from these CT studies?

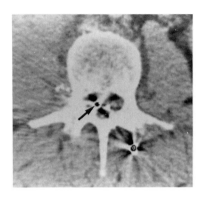 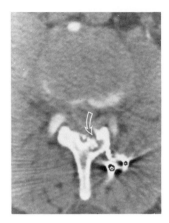 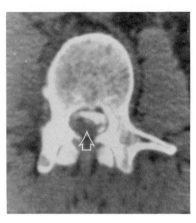

FIG. 58A-1. Bullet injury of the lumbar spine. Metallic bullet fragments are seen within the spinal canal (*arrow*) and posterolateral to the left lamina, revealing the path of the bullet. Small bone fragments are present lateral to the right pedicle and within the canal. At bone window settings, a fracture of the right pedicle could be seen. Note the artifact created by the bullet fragments.

FIG. 58B-1. This scan demonstrates fracture of the left lamina (*arrow*). Bullet fragments are again seen posteriorly.

FIG. 58C-1. CTM performed 9 days later demonstrates large hematoma (*arrow*) within the canal causing marked compression of the contrast-filled thecal sac. The hematoma measured 58 HU.

Bullet Fragments and Hematoma within Spinal Canal

There is a bullet injury of the lower lumbar spine. CT demonstrates osseous fracture as well as metallic and bony fragmentation within the spinal canal (Figs. 58A-1, 58B-1). Nine days later as the patient's clinical condition worsened, a myelogram was performed and demonstrated a block. This was followed by CTM. Transaxial scans demonstrate a large soft-tissue epidural mass, which measures 58 HU and is causing anterior displacement of the thecal sac (Fig. 58C-1). Sagittal reconstruction defines the extent of the hematoma (Fig. 58D). Surgery was performed with evacuation of the hematoma and removal of the bullet fragments.

Patients with gunshot wounds of the cervical, thoracic, or lumbar spine require radiographic evaluation to determine the presence of metallic or bony fragments within the spine, the extent of fracture, and the presence of soft-tissue injury. CT is more accurate than conventional radiography or tomography in these evaluations.[1-3] Although CT examination of intraspinal contents may be hindered by the frequent presence of artifacts created by the metal, CT has nevertheless been found to be the most accurate method of determining the presence of metallic fragments within the canal. The actual size of bullet fragments is less accurately estimated by CT because of volume averaging.[3] Although conventional tomography can delineate the configuration of the fractures that occur secondary to bullet injury, it is less accurate than CT in assessing the presence of bony fragments within the canal.[2] Soft-tissue structures are best studied by CT, and evaluation of patients with gunshot wounds may disclose pleural effusion, pulmonary hematoma, upper airway compromise, and spinal cord damage.[1,2]

A CT scan of a patient with a gunshot wound of the cervical spine is shown (Figs. 58E, 58F). The

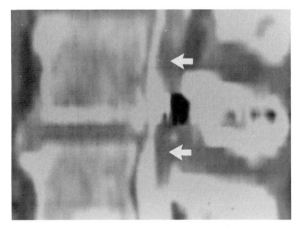

FIG. 58D. Hematoma and bullet fragments within the canal. Midsagittal reconstruction obtained from axial CTM images during the same examination as shown in Fig. 58C. Hematoma (*arrows*) is quite extensive. Despite complete myelographic block, intrathecal contrast can be seen anteriorly on the CTM study, permitting evaluation of hematoma size. In this midsagittal plane, the bullet is seen within the posterior canal.

path of the bullet can be seen traversing the vertebral body, spinal canal, and left lamina. The CT scan clearly demonstrates fractures of the vertebral body and lamina as well as bony and metallic fragments within the canal. The soft-tissue structures are also well visualized, and subcutaneous emphysema is noted. The close proximity of a bullet fragment to the trachea is well demonstrated in this patient who sustained laceration of the trachea and esophagus (Fig. 58F).

Treatment of patients with gunshot wounds of the spine is controversial. Some authorities perform surgery only for specific indications, stating that laminectomy does not influence the clinical course.[4] Others recommend immediate débridement for medically stable patients with bullet fragments in the canal.[3] Removal of metallic foreign bodies from the canal is thought to be important since myelopathy has been described as a complication in a high percentage of cases.[2] In general, patients with bullet injuries of the cauda equina have a better prognosis than those with spinal cord injury.[4]

References

1. Brant-Zawadzki M, Miller EM, Federle MP: CT in the evaluation of spine trauma. *AJR* 1981;136:369–375.
2. Plumley TF, Kilcoyne RF, Mack LA: Computed tomography in evaluation of gunshot wounds of the spine. *J Comput Assist Tomogr* 1983;7:310-312.
3. Post MJD, Green BA, Quencer RM, et al: The value of computed tomography in spinal trauma. *Spine* 1982;7:417–431.
4. Yashon D, Jane JA, White RJ: Prognosis and management of spinal cord and cauda equina bullet injuries in sixty-five civilians. *J Neurosurg* 1970;32:163–170.

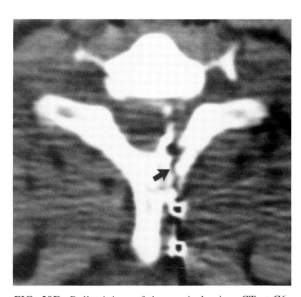

FIG. 58E. Bullet injury of the cervical spine. CT at C6. The path of bullet is seen passing through the spinal canal and the left lamina. There is fracture of the lamina (*arrow*) with fragmentation noted in the canal. Bullet fragments are located posterior to the lamina.

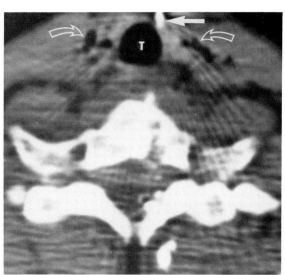

FIG. 58F. Bullet injury of the cervical spine. Same patient as in Fig. 58E. Fracture of the C7 vertebral body is seen. A metallic bullet fragment (*straight arrow*) lies in close proximity to the trachea (*T*). There is subcutaneous emphysema (*curved arrows*) on both sides of the trachea. This patient sustained laceration of the trachea and the esophagus.

CASE 59

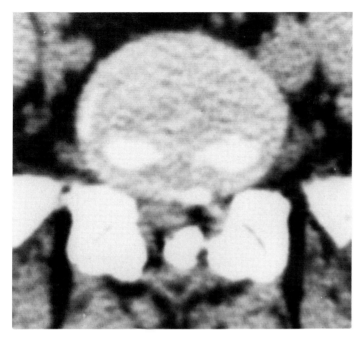

FIG. 59A. Axial CT at L5-S1 after intrathecal introduction of water-soluble contrast. This patient had had previous surgery. He was examined at this time because of acute onset of back pain and right radiculopathy, which followed 1 day after straining to lift a heavy object.

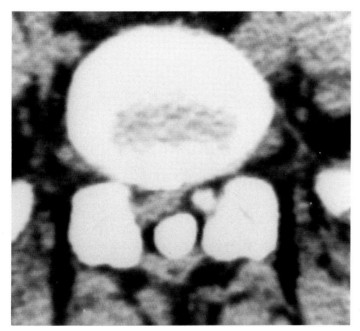

FIG. 59B. CTM 4 mm caudad to Fig. 59A.

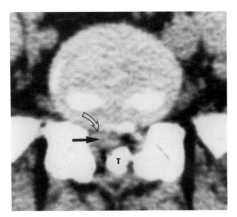
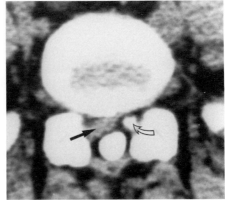

FIG. 59A-1. Free disc fragment in a postoperative patient. A large free disc fragment (*straight arrow*) fills the anterior epidural space on the right. There is partial obliteration of the anterior epidural fat and compression of the contrast-filled thecal sac. (*T*). The posteriorly displaced fragment is separated from the posterior margin of the disc by a thin layer of fat (*curved arrow*). There is a bilateral laminectomy defect.

FIG. 59B-1. The free fragment (*straight arrow*) prevents contrast from filling the right S1 nerve root sheath. Note that the normal left S1 nerve root sheath has filled with contrast (*curved arrow*).

Postoperative Free Disc Fragment

There is a soft-tissue mass in the right anterior epidural space causing compression and posterior displacement of the thecal sac (Figs. 59A-1, 59B-1). The soft-tissue mass is separated from the posterior margin of the disc by a thin layer of fat. On the following day a CT examination was performed after an intravenous injection of iodinated contrast. No enhancement of the mass is seen (Fig. 59C). These findings are due to a large free disc fragment.

Myelography is not a reliable method of distinguishing between disc herniation and fibrotic scar in the postoperative patient.[4] There is also some difficulty in making this differentiation by CT.[1,3] Both disc herniation and scar may appear as a soft-tissue density obliterating epidural fat and obscuring the margins of the thecal sac and ipsilateral nerve root. The diagnosis of a scar may be suggested if the density is linear or strandlike, contours itself around the thecal sac, or retracts the sac toward the soft-tissue density[9] (Fig. 59D). However, approximately 15% of scars have a nodular shape simulating a disc herniation.[8] Typically, a disc herniation does not follow the contour of the thecal sac and does not cause retraction of the sac, but instead compresses the thecal sac or displaces a nerve root.[9]

It is suggested that scars often extend above or below the disc space and may not appear contiguous to the posterior disc margin, whereas herniated discs are typically at the disc space level and are contiguous with the disc margin. However, an extruded disc with a free fragment is likely to extend cephalad or caudad and is separated from the posterior disc margin by fat in 50% of cases.[2,10] It is suggested that fibrosis typically has a lower density measurement than disc herniation; however, there is overlap in their range of attenuation values.

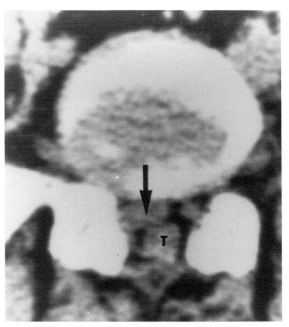

FIG. 59C. Free disc fragment. Same case as in Fig. 59A-1. CT at L5-S1 obtained on the following day with no intrathecal contrast. This study was conducted immediately after intravenous injection of iodinated contrast and reveals no enhancement of the disc fragment, further confirming the diagnosis. The free fragment (*arrow*) measures 53 HU. The thecal sac (*T*) is now seen without contrast. At surgery a large extruded free fragment was found and removed.

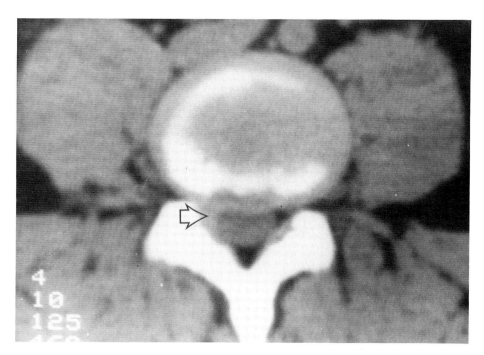

FIG. 59D. Fibrotic scar. Axial CT at the L4-L5 intervertebral disc level. This patient had had previous surgery at this level for disc herniation. A scar (*arrow*) is present on the right and contours around the thecal sac. The epidural fat is obscured by the scar; however, there is no compression of the thecal sac.

In questionable cases some authors recommend scanning with intrathecal water-soluble contrast.[6] This is especially helpful when extensive fibrosis surrounds the sac and obliterates the normal anatomic landmarks. With CTM, recurrent disc herniation typically displaces the nerve root posteriorly whereas fibrosis encompasses or causes retraction of the nerve root.[6] Another method of handling the difficult differentiation between scar and recurrent disc herniation is to inject iodinated contrast intravenously. Scar may enhance significantly whereas disc herniation does not enhance[7,8] (Fig. 59C).

In this case the patient had an acute onset of back pain and radiculopathy one day after straining to lift a heavy object. This acute onset of symptoms is more typical of disc herniation than of fibrosis. The CT findings of a soft-tissue mass in the anterior epidural space in this case favored a diagnosis of recurrent disc herniation rather than fibrosis because of the compression and displacement of the thecal sac and ipsilateral nerve root and the lack of enhancement after intravenous contrast injection. A more specific

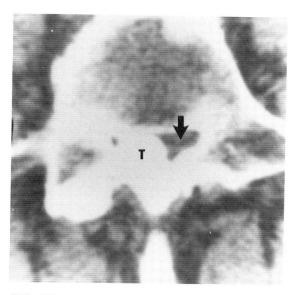

FIG. 59E. Free disc fragment in a patient with previous surgery. CTM obtained 12 mm caudad to the L4-L5 disc. The extruded disc fragment (*arrow*) is within the lateral recess on the left and displaces the contrast-filled thecal sac (*T*) to the right.

cessful whereas reoperation for fibrosis leads to further scar.[5]

In another postoperative patient, a free fragment of bone rather than disc was found (Fig. 59F). This patient had a partial left laminectomy and discectomy for disc herniation at L5-S1. In the immediate postoperative period there was increased intensity of pain, which led to further evaluation with CT. A large bone fragment was identified within the spinal canal, and reoperation was performed, again demonstrating the usefulness of CT in the postoperative setting.

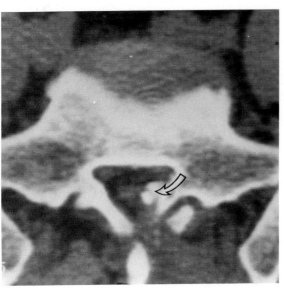

FIG. 59F. Postoperative bone fragment within the spinal canal. A partial left laminectomy and discectomy had been performed at L5-S1. Increased intensity of pain in the immediate postoperative period led to this CT study. There is a large free bone fragment within the spinal canal (*arrow*) abutting the thecal sac. An additional bone fragment is seen posterior to the remaining left lamina.

and accurate diagnosis of a free disc fragment was made by noting that the mass was separated from the posterior disc margin by a thin layer of epidural fat.

A CT scan of another postoperative patient is shown in which an epidural soft-tissue mass is seen compressing and displacing the thecal sac (Fig. 59E). This abnormality was present on additional scans over a length of approximately 2 cm. Although the long segment might suggest scarring, the mass effect is more typical of an extruded disc. A large free disc fragment was found at surgery. The distinction between scar and recurrent disc herniation cannot always be made by CT; however, when this differentiation can be made it is clinically important since reoperation for recurrent disc herniation is often suc-

References

1. Braun IF, Lin JP, Benjamin MV, et al: Computed tomography of the asymptomatic postsurgical lumbar spine: Analysis of the physiologic scar. *AJNR* 1983;4:1213–1216, *AJR* 1984;142: 149–152.
2. Dillon WP, Kaseff LG, Knackstedt VE, et al: Computed tomography and differential diagnosis of the extruded lumbar disc. *J Comput Assist Tomogr* 1983;7:969–975.
3. Haughton VM, Eldevik OP, Magnaes B, et al: A prospective comparison of computed tomography and myelography in the diagnosis of herniated lumbar disks. *Radiology* 1982;142: 103–110.
4. Irstam L: Differential diagnosis of recurrent lumbar disc herniation and postoperative deformation by myelography: An impossible task. *Spine* 1984;9:759–763.
5. Law JD, Lehman RAW, Kirsch WM: Reoperation after lumbar intervertebral disc surgery. *J Neurosurg* 1978;48:259–263.
6. Meyer JD, Latchaw RE, Roppolo HM, et al: Computed tomography and myelography of the postoperative lumbar spine. *AJNR* 1982;3:223–228.
7. Schubiger O, Valavanis A: CT differentiation between recurrent disc herniation and postoperative scar formation: The value of contrast enhancement. *Neuroradiology* 1982; 22:251–254.
8. Schubiger O, Valavanis A: Postoperative lumbar CT: Technique, results, and indications. *AJNR* 1983;4:595–597.
9. Teplick JG, Haskin ME: CT of the postoperative lumbar spine. *Radiol Clin North Am* 1983;21:395-420.
10. Williams AL, Haughton VM, Daniels DL, et al: Differential CT diagnosis of extruded nucleus pulposus. *Radiology* 1983; 148:141–148.

CASE 60

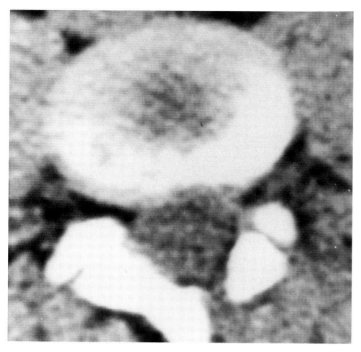

FIG. 60A. CT scan at L4-L5. There had been previous surgery at this level for disc herniation.

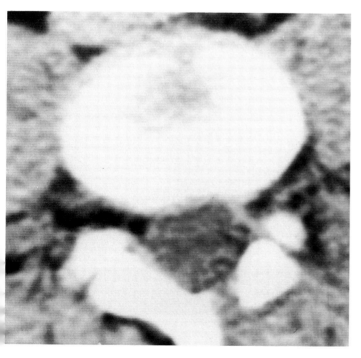

FIG. 60B. CT scan at the same level as in Fig. 60A obtained after intravenous injection of iodinated contrast.

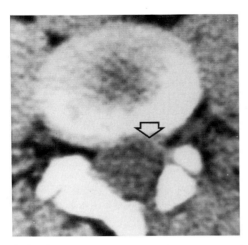 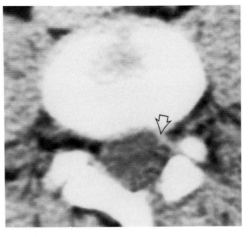

FIG. 60A-1. Precontrast scan demonstrates soft-tissue density in the left anterior epidural space (*arrow*) partially obliterating epidural fat. The differential diagnosis includes fibrotic scar and recurrent disc herniation. Note the partial laminectomy on the left.

FIG. 60B-1. Fibrotic scar. After intravenous contrast injection there is marked enhancement of the abnormal area, indicating the presence of scar (*arrow*) rather than recurrent disc herniation. The linear configuration of the scar is now more apparent.

Postoperative Fibrotic Scar: Intravenous Contrast Enhancement

There is a soft-tissue density in the left anterior epidural space obliterating the epidural fat (Fig. 60A-1). The main differential diagnosis includes recurrent disc herniation and fibrotic scar in this patient who has had previous surgery at this level. Although characteristic CT appearances of disc herniation and fibrotic scar have been described with unenhanced CT studies, the differentiation between these two entities in an individual case may be very difficult. Some authors have recommended the use of intravenous injection of iodinated contrast medium in an effort to distinguish disc material from scar.[1,3–6]

Initially, the unenhanced scan is obtained and evaluated while the patient remains on the examining table. If there is diagnostic uncertainty as to whether a CT finding represents disc material or scar at the level of previous surgery, intravenous contrast may be injected. This can be accomplished with a rapid drip infusion (3 to 5 min) of 100 to 150 ml of 60% iodinated contrast followed by repeat scanning while the contrast infusion continues,[6] or with a bolus injection of contrast at a concentration of 600 mg iodine per kilogram of body weight followed by immediate scanning.[4] The poorly vascularized disc material does not enhance, whereas scar may enhance significantly.[1,3–6] When disc herniation and scar coexist the disc material remains unenhanced, whereas the scar may be seen as an enhancing band surrounding the disc, or the scar may have a nodular appearance.[4]

When CT visualization of a nerve root sheath is obliterated by scar, the nerve root may be seen as a relative lucency after intravenous contrast enhancement of the surrounding fibrosis. In this way the position of the nerve root sheath can be evaluated, and significant displacement suggests the possibility of coexisting disc herniation.[4] Some cases, however, remain indeterminate despite the use of intravenous contrast.[1] For example, lucency within an enhancing scar could be due to either nerve root or disc fragment surrounded by scar. A definitive diagnosis may not always be possible. Nevertheless, preliminary results suggest that the use of intravenous contrast increases diagnostic confidence and accuracy.[1] The technique of rapid injection (either drip or bolus) with immediate scanning appears to be important since one report suggests that disc herniation may enhance 40 minutes after a slow drip infusion (30 minutes) of 250 ml of 60% iodinated contrast.[2]

In this case, a linear scar is enhanced after intravenous contrast injection (Fig. 60B-1). The scar extends from the lamina adjacent to the left lateral border of the dural sac to the intervertebral disc. Enhancement of a scar after intravenous contrast injection may be demonstrated many years after surgery;[4,5] however, the enhancement tends to be most intense when the scar has formed in response to a more recent operation.[4]

References

1. Braun IF, Hoffman JC Jr, Davis PC, et al: Contrast enhancement in CT differentiation between recurrent disk herniation and postoperative scar: Prospective study. *AJNR* 1985;6:607–612, *AJR* 1985;145:785–790.
2. DeSantis M, Crisi G, Folchi VF: Late contrast enhancement in the CT diagnosis of herniated lumbar disk. *Neuroradiology* 1984;26:303–307.
3. Schubiger O, Valavanis A: CT differentiation between recurrent disc herniation and postoperative scar formation: The value of contrast enhancement. *Neuroradiology* 1982;22:251–254.
4. Schubiger O, Valavanis A: Postoperative lumbar CT: Technique, results, and indications. *AJNR* 1983;4:595–597.
5. Teplick JG, Haskin ME: Computed tomography of the postoperative lumbar spine. *AJNR* 1983;4:1053–1072, *AJR* 1983;141:865–884.
6. Teplick JG, Haskin ME: Intravenous contrast-enhanced CT of the postoperative lumbar spine: Improved identification of recurrent disk herniation, scar, arachnoiditis, and diskitis. *AJNR* 1984;5:373–383, *AJR* 1984;143:845–855.

CASE 61

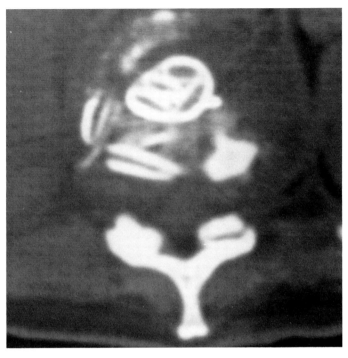

FIG. 61A. This patient had tuberculous osteomyelitis of T11-T12 with considerable osseous destruction, loss of disc space, and kyphosis. This CT scan followed shortly after surgical intervention.

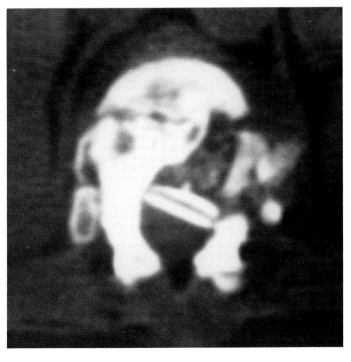

FIG. 61B. Axial CT 6 mm caudad to Fig. 61A. What postsurgical observations can be made?

 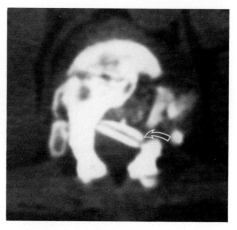

FIG. 61A-1. Vertebral body fusion with posterior displacement of the graft section. Autogenous rib fascicle bone graft has been utilized for fusion. The rib sections are seen end-on anteriorly (*straight arrow*). Another rib section was positioned posteriorly in a horizontal plane (*curved arrow*).

FIG. 61B-1. The posterior rib section is moderately displaced posteriorly on the left side (*arrow*). The patient had no neurological complications, and no further surgical intervention was required.

Surgical Fusion With Displacement of Bone Graft

This patient had tuberculous osteomyelitis with destruction of the vertebral body on both sides of the T11-T12 intervertebral disc space. Surgical intervention included an anterior vertebral fusion accomplished by the insertion of autogenous rib fascicle bone graft. The rib segments, which were wired together and placed anteriorly at the site of bone destruction, are seen end-on with transaxial CT examination (Fig. 61A-1). An additional rib section was positioned transversely in the location of the posterior portion of the vertebral body (Fig. 61B-1). Moderate displacement of this bone graft into the canal is further delineated by the sagittal reconstruction view (Fig. 61C). The patient had no neurologic compromise, and there was no further surgical intervention.

Surgical fusion may be interbody, posterior, or lateral and is performed for stabilization of various pathologic conditions such as instability related to trauma, infection, tumor, spondylolisthesis, or arthritis. Lumbar interbody fusion can be accomplished by either an anterior or a posterior surgical approach. Following the removal of disc material, bone plugs are positioned in the intervertebral disc space extending from one vertebral body to the next. Early, in the first weeks after surgery, the interbody graft is not incorporated into the matrix of the host bone, and a discrete boundary between host and graft can be seen with sagittal and coronal reformatted images.[5] As solid fusion occurs, there is obliteration of the host-graft interface. The most significant complications of interbody fusion are failure of osseous fusion with pseudoarthrosis, graft displacement, and degenerative disc disease at the level above solid fusion.[3,5] When failure of osseous fusion occurs, CT demonstrates a lack of callus and fusion between the vertebral body and the bone graft with disintegration of the graft, lucency at the involved vertebral body endplate, and loss of disc height.[3,5] This may be best visualized with reformated images. A potentially serious complication of interbody fusion is posterior displacement of a bone

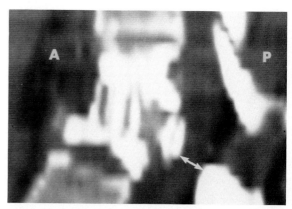

FIG. 61C. Vertebral body fusion with posterior displacement of the graft section. Sagittal reconstruction of axial images from the same patient as in Fig. 61A-1. Reconstructed images show moderately severe spinal canal compromise (*arrows*) caused by posterior displacement of the bone graft. The longitudinal extent of the anterior rib fascicle graft can be appreciated in this plane. *A*, anterior; *P*, posterior.

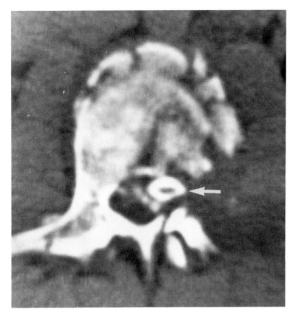

FIG. 61D. Bone plug with posterior displacement into the spinal canal. This patient was treated for an unstable burst fracture of L3. Postoperatively the bone plug can be seen within the spinal canal (*arrow*).

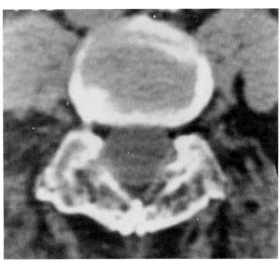

FIG. 61E. Solid posterolateral fusion. Axial CT demonstrates solid posterolateral lumbar fusion and ankylosis of joints. There is no osseous encroachment on the spinal canal at this level.

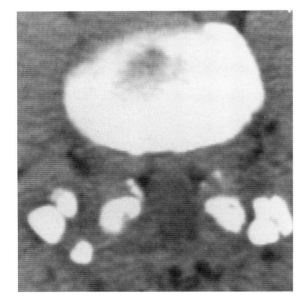

FIG. 61F. Incomplete fusion. This patient had had an attempted fusion at L4-L5 1 year earlier. Multiple small fragments of bone are seen without evidence of solid fusion. A wide bilateral laminectomy is noted.

plug. The bone plug may be displaced into the spinal or neural canal and can be seen on the transaxial CT scan extending posterior to the adjacent vertebral bodies (Fig. 61D). Sagittal reformatting may be helpful in determining whether the plug is truly displaced into the canal. Posterior displacement of a bone plug can be associated with a sudden onset of symptoms and may require reoperation.

The more commonly performed posterior lumbar fusion is achieved with bone graft placed between the laminae and spinous processes of adjacent vertebrae. Lateral fusion is accomplished with bone graft placed between transverse processes. CT can be used to evaluate the solidity of a posterior or lateral fusion as well as its complications.[6] Immediately after surgery, discrete isolated bone is normally demonstrated. In time solid fusion occurs, with increased bone mass encompassing the laminae, spinous processes, and articular processes. Ankylosis of facet joints may develop (Fig. 61E). When there is a failure of fusion, multiple fragments of bone graft remain isolated without coalescing (Fig. 61F). Pseudoarthrosis and incomplete facet joint fusion can be identified. Spinal stenosis may develop after surgical fusion especially when posterior fusion has been performed. Stenosis most often occurs at the disc space level immediately above the superior extent of fusion and is due to a combination of disc herniation, thickening and infolding of the ligamenta flava, medial hypertrophy of articular processes, and ventral projection of the upper margin of the fusion mass.[1,4] Less frequently, diffuse stenosis may occur at the level of fusion and is due to hypertrophy of the midline fusion mass causing neural encroachment.[1,4]

Cervical fusion can also be evaluated by CT. Anterior cervical fusions have a high complication rate thought to be related to performance of the anterior fusion in the presence of unrecognized posterior instability.[2] The most frequent complications are extrusion of bone graft and kyphosis.[2]

References

1. Brodsky AE. Post-laminectomy and post-fusion stenosis of the lumbar spine. *Clin Orthop* 1976;115:130–139.
2. Foley MJ, Lee C, Calenoff L, et al: Radiologic evaluation of surgical cervical spine fusion. *AJR* 1982;138:79–89.
3. Golimbu C, Firooznia H, Rafii M, et al: Computed tomography of thoracic and lumbar spine fractures that have been treated with Harrington instrumentation. *Radiology* 1984;151:731–733.
4. Grabias S: The treatment of spinal stenosis. *J Bone Joint Surg Am* 1980;62-A:308–313.
5. Rothman SLG, Glenn WV Jr: CT evaluation of interbody fusion. *Clin Orthop* 1985;193:47–56.
6. Teplick JG, Haskin ME: Computed tomography of the postoperative lumbar spine. *AJNR* 1983;4:1053–1072, *AJR* 1983;141:865–884.

CASE 62

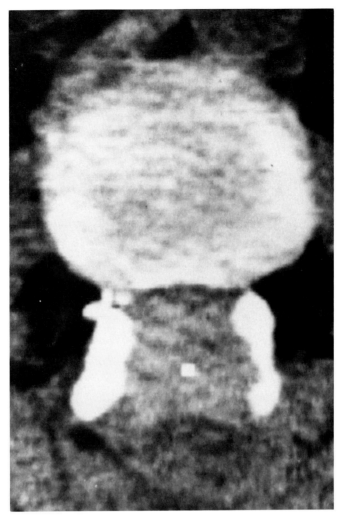

FIG. 62A. Axial CT at L1–L2. This 60-year-old had spinal surgery with decompressive laminectomy for spinal stenosis. Postoperative strength and sensation were initially normal. The CT scan was obtained on the third postoperative day after he developed weakness in both legs and inability to void. The white dot in the center of the spinal canal marks the site of a CT density measurement reading of 73 HU.

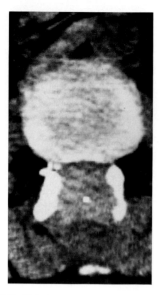

FIG. 62A-1. Acute epidural hematoma. There is increased density within the entire canal, measuring 73 HU. This density measurement is significantly greater than that normally found within the thecal sac. This represents a large epidural hematoma. Immediate surgical evacuation of the hematoma was performed with subsequent return of strength and function.

Epidural Hematoma

The CT scan reveals a large area of increased density involving the entire spinal canal at this level (Fig. 62A-1). The soft tissue within the canal measures 73 HU and is certainly of much greater density than would be expected in a normal thecal sac. The CT findings along with the clinical history suggest the diagnosis of a large acute epidural hematoma in this postoperative patient. Reoperation confirmed this diagnosis.

Epidural hematoma may occur after spinal surgery or lumbar puncture, or in association with trauma, infection, neoplasm, coagulopathy, or anticoagulant therapy.[6,7] Occasionally epidural hematoma may occur spontaneously.[2] Clinically an acute epidural hematoma represents an emergency that can lead to severe neurologic complications and possible death if not treated promptly. Patients present with severe back or neck pain, which occurs suddenly and is often radicular. This may progress to weakness and urinary retention and then lead to paraplegia or quadriplegia.[6] Epidural hematoma has been found in approximately 0.2% of patients who have had spinal surgery for lumbar disc herniation.[1] These patients may have an immediate symptom-free postoperative period lasting approximately 1 day, which is followed by persistent

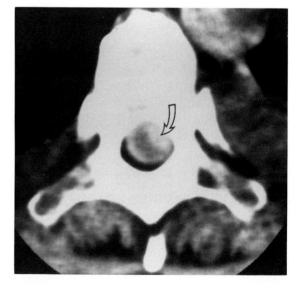

FIG. 62B. Acute epidural hematoma and subarachnoid hemorrhage. This patient had a history of anticoagulant therapy and experienced bilateral lower extremity weakness. The CT study of the thoracic spine demonstrates a lenticular, high-density epidural hematoma (*arrow*) causing compression of the spinal cord. Hemorrhage is also noted in the subarachnoid space. (Figure courtesy of Jeffrey Blinder, M.D., Allentown, Pa.)

painless paresthesias over the dermatome of the exposed nerve root.[1]

The CT examination can be an important part of the diagnostic workup in a patient in whom epidural hematoma is suspected.[2,3,6,7] CT can be used as the initial imaging modality when the clinical presentation suggests an epidural hematoma and signs and symptoms permit spinal level localization. However, when a clinical level cannot be determined, myelography may precede CTM.[4] In the acute stage, the CT study performed without intrathecal contrast reveals a peripheral, sharply defined biconvex mass of increased density (approximately 60 to 80 HU) (Fig. 62B) or an area of similarly increased density filling the entire canal[2,5-7] (Fig. 62A-1). Frequently the hematoma is several centimeters in length and extends beyond a single vertebral level. If diagnostic uncertainty remains, myelography with water-soluble contrast followed by CTM can be performed. With CTM the epidural hematoma appears as a lenticular mass compressing the contrast-filled subarachnoid space and displacing the spinal cord or cauda equina.[3,4,6,7] In the subacute stage, an epidural hematoma may be isodense with the thecal sac and is further delineated by CTM.[7] Similar CT findings have been demonstrated with the clinically more unusual chronic epidural hematoma.[5] Sagittal reconstruction of a chronic epidural hematoma may reveal tapered margins, unlike a tumor mass.[5]

References

1. DiLauro L, Poli R, Bortoluzzi M, et al: Paresthesias after lumbar disc removal and their relationship to epidural hematoma. *J Neurosurg* 1982;57:135–136.
2. Haykal HA, Wang A-M, Zamani AA, et al: Computed tomography of spontaneous acute cervical epidural hematoma. *J Comput Assist Tomogr* 1984;8:229–231.
3. Kaiser MC, Capesius P, Ohanna F, et al: Computed tomography of acute spinal epidural hematoma associated with cervical root avulsion. *J Comput Assist Tomogr* 1984;8:322–323.
4. Lanzieri CF, Sacher M, Solodnik P, et al: CT myelography of spontaneous spinal epidural hematoma: Case report. *J Comput Assist Tomogr* 1985;9:393–394.
5. Levitan LH, Wiens CW: Chronic lumbar extradural hematoma: CT findings. *Radiology* 1983;148:707–708.
6. Post MJD, Seminer DS, Quencer RM: CT diagnosis of spinal epidural hematoma. *AJNR* 1982;3:190–192.
7. Zilkha A, Irwin GAL, Fagelman D: Computed tomography of spinal epidural hematoma. *AJNR* 1983;4:1073–1076.

CASE 63

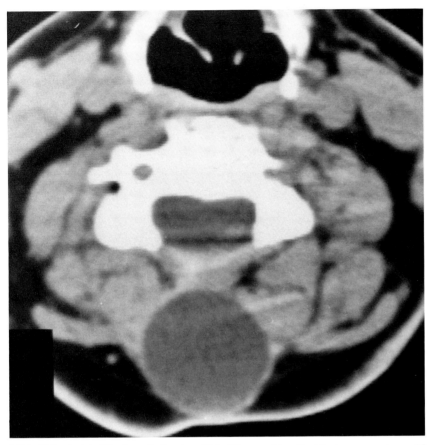

FIG. 63A. Axial CT of cervical spine. This 33-year old patient had surgery at age 15 for an astrocytoma. What postsurgical finding is demonstrated?

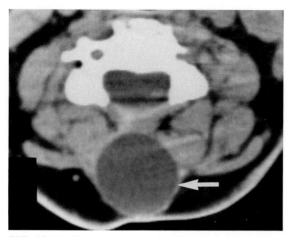

FIG. 63A-1. Pseudomeningocele. There is a well-circumscribed, encapsulated, low-density "mass" (arrow) posterior to the spinal canal at the site of previous bilateral laminectomy. This is a postoperative pseudomeningocele. The pseudomeningocele measured 3.5 cm in diameter and had density measurements of 5 HU.

Pseudomeningocele

There is a large, round, well-circumscribed low-density mass posterior to the thecal sac in this patient who has had a previous laminectomy (Fig. 63A-1). This is a pseudomeningocele, which is a posteriorly localized collection of cerebrospinal fluid usually resulting from an inadequate closure or inadvertent tear of the dura following laminectomy.[3-5] It may also develop following lumbar puncture or after avulsion of cervical spinal nerve roots.[1,3] Usually pseudomeningoceles form from leakage of cerebrospinal fluid, which eventually becomes surrounded by a fibrous capsule. Less often, intact arachnoid may herniate through the tear, and enlargement of an arachnoid-lined true meningocele may occur.[5] Pseudomeningocele may present from 1 month to several years following laminectomy. It is most frequent in the lower lumbar spine and least frequent in the thoracic spine. Approximately 2% of symptomatic postoperative patients have pseudomeningocele; however, the relationship between the pseudomeningocele and symptoms is uncertain.[5]

The diagnosis of a pseudomeningocele can be accurately made with CT.[3-5] It appears as a round, homogeneous mass of low CT attenuation similar to cerebrospinal fluid. In the lumbar region pseudomeningocele is similar or slightly lower in density compared to the thecal sac, whereas in the cervical region it has significantly lower attenuation values than the thecal sac, probably because of the presence of higher density spinal cord within the sac.[5] Pseudomeningoceles vary in size, are usually unilocular, and develop posterior to the sac at the laminectomy site.[2,5] Frequently the pseudomeningocele appears to be partly or completely contained within a higher density capsule that occasionally calcifies.[1,4,5]

Conventional myelography may sometimes fail to demonstrate a pseudomeningocele.[3,4] CTM may be helpful by demonstrating even low concentrations of contrast in the pseudomeningocele that were not visualized with conventional myelography.[3,4] Some authors recommend CTM scanning of patients with pseudomeningocele in both the prone and the supine position to demonstrate the exact site of communication with the subarachnoid space.[3] This additional information may be helpful to the surgeon contemplating repair of the dural tear. It should be remembered that, following laminectomy in the normal postoperative patient, the thecal sac may sometimes appear to bulge slightly posteriorly. The normal posterior bulging of an intact thecal sac should not be confused with a pseudomeningocele.[5] In addition to CT, sonography has been utilized to demonstrate a pseudomeningocele in patients with previous laminectomy.[2]

References

1. Carollo C, Rigobello L, Carteri A, et al: Postsurgical calcified pseudocyst of the lumbar spine. *J Comput Assist Tomogr* 1982;6:627–629.
2. Laffey PA, Kricun ME: Sonographic recognition of postoperative meningocele. *AJNR* 1984;5:329–330, *AJR* 1984;143:177–178.
3. Patronas NJ, Jafar J, Brown F: Pseudomeningoceles diagnosed by metrizamide myelography and computerized tomography. *Surg Neurol* 1981;16:188–191.
4. Ramsey RG, Penn RD: Computed tomography of a false postoperative meningocele. *AJNR* 1984;5:326–328.
5. Teplick JG, Peyster RG, Teplick SK, et al: CT identification of postlaminectomy pseudomeningocele. *AJNR* 1983;4:179–182. *AJR* 1983;140:1203–1206.

CASE 64

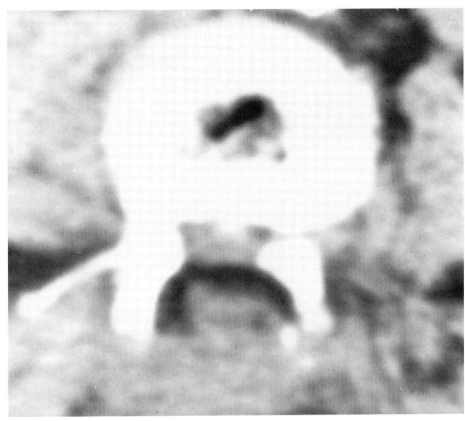

FIG. 64A. CT scan at L3-L4. This 74-year-old female had had surgery for lumbar stenosis 1 month prior to this CT examination. What operative technique can be evaluated on this scan?

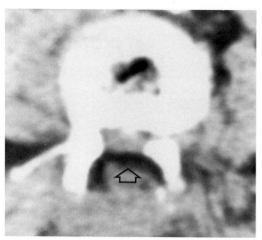

FIG. 64A-1. A fat graft (*arrow*) has been placed within the posterior spinal canal and is readily apparent by virtue of its negative CT attenuation values. Early in the postoperative stage the fat appears to compress the thecal sac. Note also the wide bilateral laminectomy defect and the vacuum disc phenomenon.

Surgical Fat Graft

A free fat graft and bilateral decompressive laminectomy have been performed for the treatment of this patient's lumbar stenosis (Fig. 64A-1). The intraoperative use of free and pedicle fat grafts in patients treated surgically for spinal stenosis has been recommended as a means of preventing postoperative fibrosis.[2-4,6] Large quantities of fat, 1 cm or more in thickness, are positioned on the dura and may also be placed around the nerve roots in the lateral recess. It is thought that the fat acts as a mechanical barrier preventing serous fluid and blood from collecting at the surgical site and thus inhibiting the formation of epidural and perineural fibrosis.[3,6] Scar formation has been prevented for months to years in both clinical trials and animal experiments.[4] Although fat grafts inhibit epidural fibrosis, the exact relationship between fibrosis and clinical symptomatology is not clear since fibrosis is not infrequently seen in relatively asymptomatic patients as well as those with the failed back surgery syndrome.[1]

CT clearly demonstrates the presence and extent of the fat graft, which has a negative attenuation value. In the early postoperative period the fat graft compresses the thecal sac, as in this case. This compression of the thecal sac does not appear to be a cause of symptoms. More than 6 months after surgery the fat volume decreases to approximately 30% to 50% of the original size, and compression of the thecal sac abates.[2,3] The fat graft can still be demonstrated by CT more than 15 years after surgery.[5] When a free fat graft is used, CT demonstration of the fat is seen in more than 60% of cases studied postoperatively and indicates that revascularization of the free fat graft has occurred.[2] It is possible that the lack of CT evidence of a fat graft in a patient in whom a free fat graft was used may suggest that revascularization has not taken place.[2]

The CT evaluation of the postoperative patient may be limited by extensive fibrosis, which distorts normal landmarks. A secondary benefit of the use of intraoperative fat grafts is that the presence of fat around the thecal sac makes the CT examination of the postoperative patient easier to interpret.[3]

References

1. Braun IF, Lin JP, Benjamin MV, et al: Computed tomography of the asymptomatic postsurgical lumbar spine: Analysis of the physiologic scar. *AJNR* 1983;4:1213–1216, *AJR* 1984;142: 149–152.
2. Bryant MS, Bremer AM, Nguyen TQ: Autogenic fat transplants in the epidural space in routine lumbar spine surgery. *Neurosurgery* 1983;13:367–370.
3. Heithoff KB: High-resolution computed tomography and stenosis: An evaluation of the causes and cures of the failed back surgery syndrome, in Post MJD (ed): *Computed Tomography of the Spine*. Baltimore, Williams & Wilkins, 1984, pp 506–545.
4. Langenskiöld A, Kiviluoto O: Prevention of epidural scar formation after operations on the lumbar spine by means of free fat transplants: A preliminary report. *Clin Orthop* 1976;115: 92–95.
5. Langenskiöld A, Valle M: Epidurally placed free fat grafts visualized by CT scanning 15–18 years after discectomy. *Spine* 1985;10:97–98.
6. Yong-Hing K, Reilly J, deKorompay V, et al: Prevention of nerve root adhesions after laminectomy. *Spine* 1980;5:59–64.

CASE 65

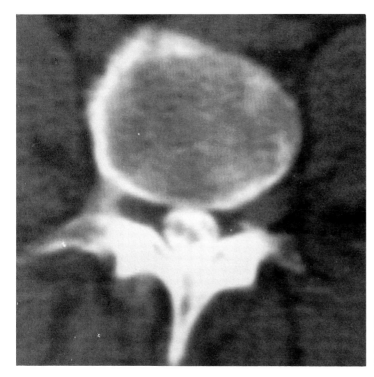

FIG. 65A. CTM at L4. This 53-year-old female had had previous lumbar surgery and now has recurrent pain in the lumbosacral region.

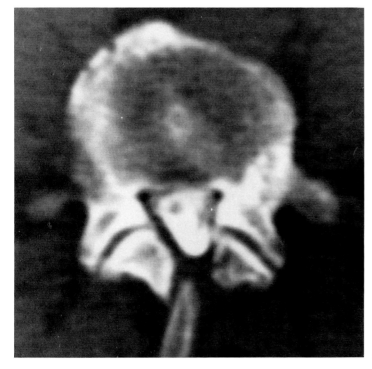

FIG. 65B. CTM at L5.

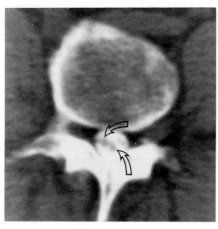 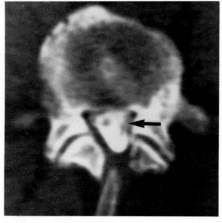

FIG. 65A-1. Arachnoiditis. The nerve roots of the cauda equina are visualized within the contrast-filled thecal sac. The nerve roots appear thickened and are clumped together (*arrows*).

FIG. 65B-1. The nerve roots of the cauda equina are thickened and matted peripherally and adherent to the dural margins (*arrow*).

Arachnoiditis

This CTM study demonstrates thickening and peripheral clumping of the nerve roots of the cauda equina (Figs. 65A-1, 65B-1). These findings are due to arachnoiditis. The nerve roots of the cauda equina are best visualized when the CTM study is viewed at bone window settings (e.g., 1,000 to 2,000 HU). In the axial plane, normal nerve roots of the cauda equina appear as multiple small, round filling defects evenly distributed throughout the thecal sac (Figs. 65C, 65D). CTM examination of arachnoiditis may demonstrate clumping of the nerve roots of the cauda equina, or peripheral adherence of the nerve roots to the dural margins leading to an "empty thecal sac" appearance.[11,12] With extensive arachnoiditis nerve roots may form a single tubular mass (Fig. 65E).

Arachnoiditis is a noninfectious inflammation of all the meningeal layers: the pia, the arachnoid, and the dura. It is thought to be the primary pathologic process in 6% to 16% of all patients with the failed back surgery syndrome.[5] Patients with arachnoiditis almost invariably have low back and leg pain that is increased by activity. Often, motor, sensory, and reflex deficits are bilateral and/or multilevel.[2,4] Urinary and bowel sphincter dysfunction may be present. Arachnoiditis most frequently occurs in patients who have had a history of disc herniation[10] and have had previous surgery and Pantopaque myelography.[2,8] It is thought that the disc herniation acts as a primary inflammatory focus, which is then potentiated by an extrinsic process such as surgery or Pantopaque contrast.[5] Other causes of arachnoiditis include spinal trauma, infection, tumor, hemorrhage, spinal anesthesia, and intrathecal serum injection.[2,8] Some cases are idiopathic.

Myelography is the initial imaging procedure in patients thought to have arachnoiditis and may be diagnostic if the radiographic and clinical findings are typical. However, CTM may provide important additional information especially in patients who have a myelographic block or an inconclusive myelogram[11] (Figs. 65F–65I). In one series, the myelographic diagnosis of arachnoiditis was frequently found to have coexistent pathology such as spinal stenosis, foraminal nerve root entrapment, and disc herniation, which, when surgically corrected, often resulted in restoration of normal function.[3] CT may play an important role in diagnosing such coexistent pathology.

CT can also be used to evaluate the sequelae of arachnoiditis. For example, patients with arachnoiditis may develop intradural arachnoid cysts or intramedullary cavities (syringomyelia).[8] Filling of the cystic structure may be demonstrated on CT scans obtained 6 to 24 hours after introduction of intrathecal water-soluble contrast.[11] A late sequela of adhesive arachnoiditis is arachnoiditis ossificans, a proliferative bony metaplasia of the arachnoid that closely envelops the spinal cord and nerve roots.[7] Unenhanced CT may reveal either (1) a thin, circumferential ring of calcification or ossification surrounding the arachnoid or (2) a large, thick, tubular bony mass[1,6,9] (Figs. 65J, 65K). This is a rare entity, which differs from the small, benign, calcific arachnoid plaques frequently reported at autopsy. Etiologic factors that have been implicated include vascular anom-

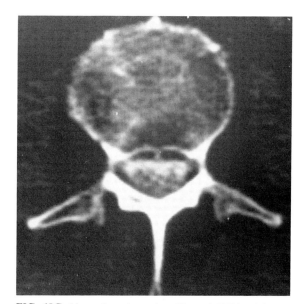

FIG. 65C. Normal cauda equina. Normal CTM at L3. The nerve roots of the cauda equina appear as multiple small, round, uniform filling defects within the contrast-filled thecal sac. Bone window settings are needed to best visualize these nerve roots

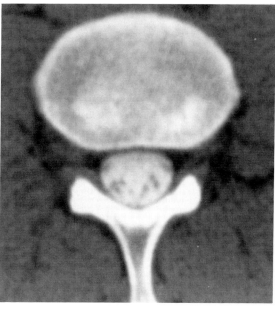

FIG. 65D. Normal cauda equina. CTM at L4-L5 in another patient. The nerve roots are less numerous and more peripheral than those shown in Fig. 65C. Nevertheless, the roots are small and rounded and are symmetrically positioned within the secal sac.

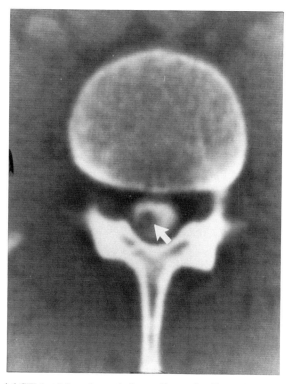

FIG. 65E. Arachnoiditis. Axial CTM at L3 performed after a "complete" myelographic block. CTM study demonstrates intrathecal contrast beyond the block. The nerve roots are matted together into a single tubular mass (*arrow*). This patient had had surgery 3 years earlier for disc disease.

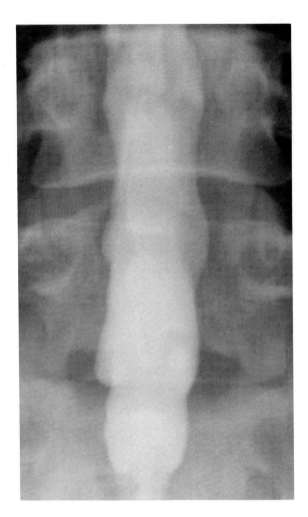

FIG. 65F. Arachnoiditis. AP radiograph of the lumbar spine from a myelogram performed with intrathecal water-soluble contrast. There is bilateral uniform lack of filling of the nerve root sleeves.

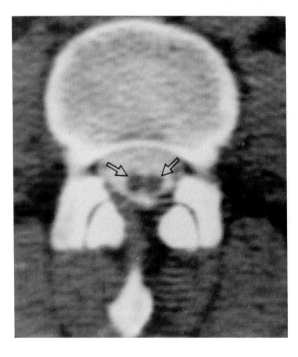

FIG 65G. Arachnoiditis. Same patient as in Fig. 65F. Axial CTM demonstrates marked clumping of the nerve roots (*arrows*).

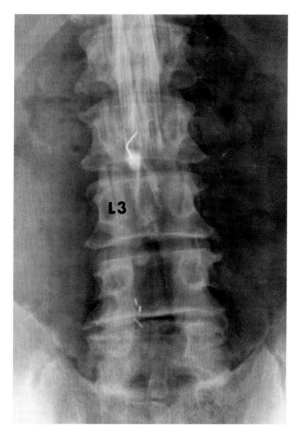

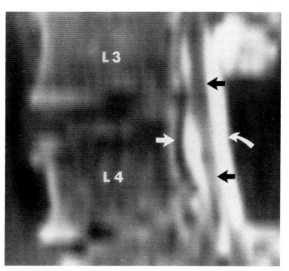

FIG. 65I. Arachnoiditis. Same patient as in Fig. 65H. Sagittal reconstruction of axial images demonstrates intrathecal contrast below the level of the block (*curved arrow*). The nerve roots of the cauda equina are thickened and clumped together (*straight arrows*). Note the absence of posterior elements at L4.

FIG. 65H. Arachnoiditis. This 60-year-old male had a history of previous Pantopaque myelography and lumbar laminectomy 2 years earlier. He had a myelogram at this time because of bilateral leg numbness, tingling, and burning dysesthesias. A "complete" myelographic block was present at L2-L3. CTM (shown in Fig. 65I) was then performed for further evaluation.

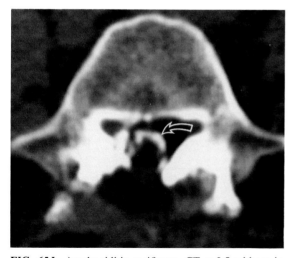

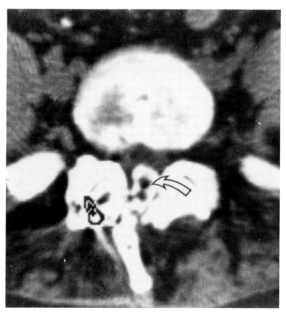

FIG. 65J. Arachnoiditis ossificans. CT at L5 without intrathecal contrast. There is curvilinear ossification of the arachnoid (*arrow*). Lateral recess stenosis is noted on the right. This patient had had previous Pantopaque myelography and spinal surgery. Note the laminectomy defect.

FIG. 65K. Arachnoiditis ossificans. Same patient as in Fig. 65J. CT without intrathecal contrast. There is a large bony mass within the spinal canal. This ossified mass has filling defects within it, which proved at surgery to be nerve roots of the cauda equina (*arrow*) entrapped within the bone. Note the surgical fixation screw.

alies of the spine, repeated spinal anesthesia, and previous history of meningitis, surgery, and trauma.[1,7]

Reference

1. Barthelemy CR: Case report: Arachnoiditis ossificans. *J. Comput Assist Tomogr* 1982;6:809–811.
2. Benner B, Ehni G: Spinal arachnoiditis: The postoperative variety in particular. *Spine* 1978;3:40–44.
3. Brodsky AE: Cauda equina arachnoiditis: A correlative clinical and roentgenologic study. *Spine* 1978;3:51–60.
4. Burton CV: Lumbosacral arachnoiditis. *Spine* 1978;3:24–30.
5. Burton CV, Kirkaldy-Willis WH, Yong-Hing K, et al: Causes of failure of surgery on the lumbar spine. *Clin Orthop* 1981;157:191–199.
6. Dennis MD, Altschuler E, Glenn W, et al: Arachnoiditis ossificans: A case report diagnosed with computerized axial tomography. *Spine* 1983;8:115–117.
7. Nainkin L: Arachnoiditis ossificans: Report of a case. *Spine* 1978;3:83–86.
8. Quencer RM, Tenner M, Rothman L: The postoperative myelogram: Radiographic evaluation of arachnoiditis and dural/arachnoidal tears. *Radiology* 1977;123:667–679.
9. Sefczek RJ, Deeb ZL: Case report: Computed tomography findings in spinal arachnoiditis ossificans. *CT* 1983;7:315–318.
10. Shaw MDM, Russell JA, Grossart KW: The changing pattern of spinal arachnoiditis. *J. Neurol Neurosurg Psychiatry* 1978;41:97–107.
11. Simmons JD, Newton TH: Arachnoiditis, in Newton TH, Potts DG (eds): *Computed Tomography of the Spine and Spinal Cord*. San Anselmo, Calif, Clavadel Press, 1983; pp 223–229.
12. Yeates AE, Newton TH: Applications of metrizamide in computed tomographic examination of the lumbar spine, in Genant HK, Chafetz N, Helms CA (eds): *Computed Tomography of the Lumbar Spine*. San Francisco, University of California, 1982; pp 67–86.

CASE 66

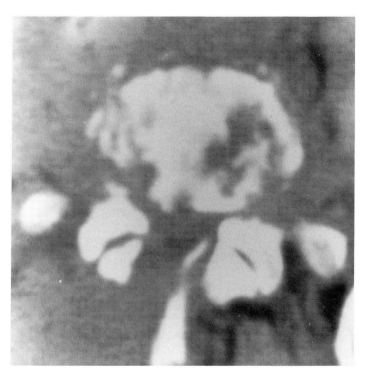

FIG 66A. CT at the inferior aspect of L4. This 57-year-old male had had surgery 2 months earlier for lumbar disc disease at L4-L5. He now presented with increasingly severe buttock and left leg pain along with paravertebral muscle spasm.

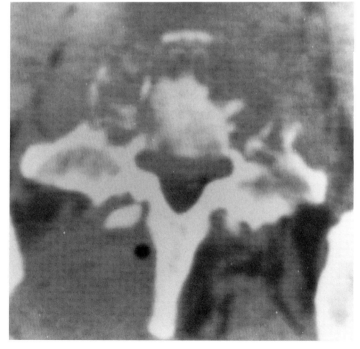

FIG. 66B. CT at the superior aspect of L5.

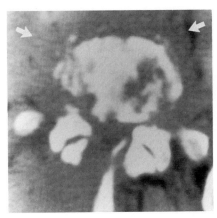 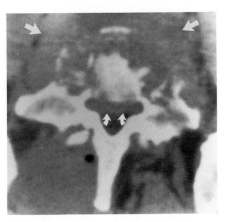

FIG. 66A-1. Postoperative pyogenic osteomyelitis. There is destruction of the inferior vertebral endplate of L4 along with marked paravertebral extension of infection (*arrows*). Postoperative changes are seen. Note the soft-tissue fullness involving the right posterior spinal structures in this patient, who had persistent drainage at the operative site.

FIG. 66B-1. Destruction of the superior vertebral endplate of L5 with paraspinal (*straight arrows*) and epidural (*curved arrows*) extension of infection. Culture revealed *Staphylococcus aureus*. Gas can be seen within the posterior paraspinal structures on the right at the site of previous surgery.

Pyogenic Osteomyelitis

This CT examination reveals changes of postoperative osteomyelitis with osteolysis of the inferior endplate of L4 and the superior endplate of L5 as well as paravertebral and epidural extension of the infection (Figs. 66A-1, 66B-1). Disc space infection occurs in 0.2% to 3.0% of patients in the postoperative period following surgery for lumbar disc disease.[10] These patients have a symptom-free early postoperative period. Symptoms usually begin between the first and fourth postoperative weeks but may sometimes present as long as 8 months after surgery.[10] Severe low back pain, which may radiate into the buttocks or legs, is the most frequent presenting symptom and is often accompanied by paravertebral muscle spasm. The presence of fever is variable. The white blood cell count is elevated in 30% of these patients.[10] The erythrocyte sedimentation rate is typically elevated at the time of surgery but returns to normal within 3 to 4 weeks after uncomplicated surgery.[10] Patients with postoperative disc space infection almost invariably have an elevated erythrocyte sedimentation rate[10] (Fig. 66C).

The postoperative state is not the most frequent clinical setting for pyogenic osteomyelitis. Most often pyogenic osteomyelitis occurs from hematogenous spread of infection. Approximately two thirds of cases of pyogenic osteomyelitis have a known source of infection, with about one half having a history of genitourinary tract infection or instrumentation.[11] Other sources of spinal osteomyelitis include soft-tissue infections, respiratory tract infections, intravenous drug abuse, and infected intravenous sites.[11]

Staphylococcus aureus is the most frequent organism causing spinal osteomyelitis and is found in over 50% of cases.[11] Other pyogenic organisms include *Escherichia coli, Proteus, Pseudomonas,* and

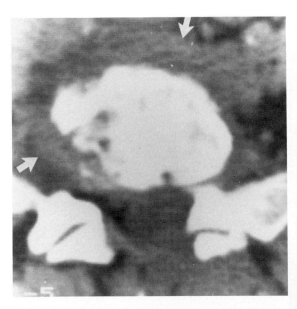

FIG. 66C. Postoperative pyogenic osteomyelitis at L5-S1. CT scan at the inferior aspect of L5. This 43-year-old woman had increasing back pain on the right side 3 months after bilateral laminectomy at this level. The patient's white blood cell count was normal whereas the erythrocyte sedimentation rate was elevated. This CT scan demonstrates multiple round, lucent lesions of the vertebral endplate, which are more marked on the right. There is paravertebral soft-tissue extension of the infection (*arrows*). Culture revealed *E. coli.*

group A *Streptococcus*. *Pseudomonas* is particularly common in heroin addicts. Hematogenous spread of infection usually begins in the vertebral body just beneath the vertebral endplate and may spread through vascular communications to involve a contiguous vertebral body, with subsequent narrowing of the less vascular intervertebral disc. In addition to the vertebral body, spinal infection may primarily involve the epidural space (epidural abscess), the disc (discitis), or the paraspinal soft tissues (paraspinal abscess).

Radionuclide scanning is a sensitive but nonspecific method of detecting early hematogenous spinal osteomyelitis. Despite its nonspecificity, radionuclide scanning is an important modality in localizing the abnormal site. Conventional radiography is typically normal in the first 2 to 8 weeks after the onset of clinical symptoms.[8] The earliest radiographic change is blurring of the vertebral endplate. Erosions of the vertebral endplates on both sides of a narrowed disc are then seen. As the infection continues, further osteolytic destruction and vertebral collapse develop.

CT, however, is considered the study of choice in evaluating patients suspected of spinal infection.[2,3,12] Experimental[9] and clinical[2,3,8,12] work suggests that osteomyelitis can be detected earlier with CT than with conventional radiography. In some cases

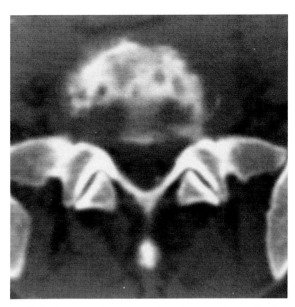

FIG. 66E. Pyogenic osteomyelitis. Same patient as in Fig. 66D. CT at L5-S1 studied at bone window settings clearly demonstrates multiple round lucencies within the inferior endplate of L5 due to infection.

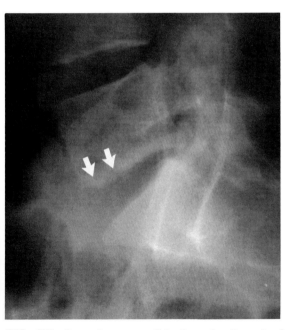

FIG. 66D. Pyogenic osteomyelitis. Lateral radiograph of the lower lumbar spine. The inferior endplate of the L5 vertebral body is indistinct anteriorly (*arrows*), suggesting the diagnosis of osteomyelitis. This patient had been evaluated for 3 weeks at another hospital for fever and back pain.

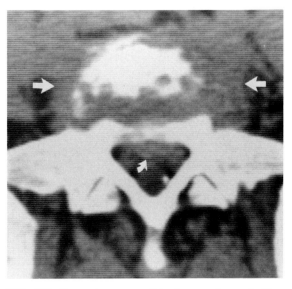

FIG. 66F. Pyogenic osteomyelitis. Same patient as in Fig. 66D. CT at L5-S1 studied at soft-tissue window settings. Paravertebral (*straight arrows*) and epidural (*curved arrow*) extension of the infection is demonstrated. This type of soft-tissue detail is best visualized by CT. Culture revealed *Staphylococcus aureus*.

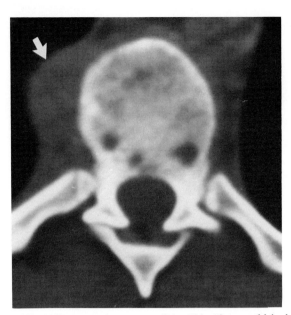
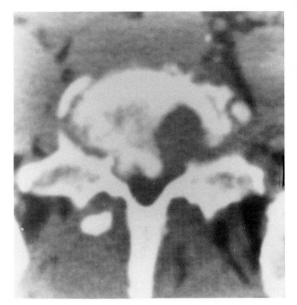

FIG. 66G. Pyogenic osteomyelitis. This 19-year-old had back pain and fever, which followed treatment for pneumonia. CT demonstrates osteomyelitis at T8-T9 with a paravertebral abscess (*arrow*). Multiple round osteolytic lesions were seen on both sides of a narrowed disc space.

FIG 66H. Postoperative pyogenic osteomyelitis with spinal stenosis. Same patient as in Fig. 66A-1. This patient had surgical intervention following the initial diagnosis of postoperative osteomyelitis. This follow-up study demonstrates decreased AP diameter of the spinal canal due to continued disc space narrowing and vertebral body subluxation. The surgical defect is noted on the left.

extensive destruction of cortical and medullary bone is detected by CT scanning while either no abnormality or only subtle erosions are seen with conventional radiography (Figs. 66D–66F). In addition to osseous changes, soft-tissue extension of infection is more readily detected by CT than by other modalities.[1–4,8,12] Paravertebral abscess or granulation tissue appears as a soft-tissue mass or swelling obliterating the normal paravertebral fat (Figs. 66A-1, 66B-1, 66G). Gas may be seen within a paravertebral abscess. Hypodensity of the lumbar disc has been described in patients with infection[7] but has not been observed by other investigators.[6] Most often, the diagnosis of osteomyelitis is made with CT when both fragmentation of the vertebral endplates and anterior paravertebral soft-tissue swelling are present.[6] Detection of a paravertebral abscess further suggests the diagnosis. False-negative studies have been reported as due to scanning at the wrong level. This may be avoided by localizing the suspected disc space by means of clinical data along with conventional radiographs and/or radionuclide bone scans obtained prior to CT examination. The CT findings of osteomyelitis in the postoperative patient are similar to those found in hematogenous osteomyelitis. However, a postdiscectomy patient may have paravertebral edema or hemorrhage, which may simulate infection.[6] Also, aggressive curettage of the vertebral endplates during discectomy may mimic the vertebral endplate erosions of osteomyelitis.[6] Nevertheless, in cases of postoperative spinal osteomyelitis, the combination of clinical data and typical bone and soft-tissue CT abnormalities should lead to the correct diagnosis.

CT examination of the spinal cord may demonstrate epidural infection with soft-tissue mass obliterating the epidural fat. The epidural inflammatory process may be chronic and represent granulation tissue associated with adjacent osteomyelitis. On the other hand, this epidural inflammation may represent an acute epidural abscess, which can occur with or without osseous involvement. Typically, an epidural abscess appears on CT as a low-density collection, which demonstrates rim enhancement after intravenous contrast injection.[12] The use of intravenous contrast to enhance the rim of the inflammatory mass provides additional accuracy in the diagnosis of epidural infection and further delineates the degree of compression of the spinal cord or thecal sac.[12] Bone fragments displaced by infection may also be detected within the canal.[4] Myelography is generally not necessary in the evaluation of osteomyelitis of the spine except in evaluating complications such as spinal cord compression with possible block. When an epidural abscess is suspected, myelography can be used to determine the level and extent of the epidural mass. The use of CTM with small-volume, low-concentration, water-soluble contrast may replace conventional myelography in the evaluation of these patients.[1]

CT scanning can aid in the follow-up examination of the patient with spinal infection by determining the presence of further bone destruction, bone fragmentation within the canal, or changes in the soft-tissue extension (Fig. 66H). Initially, patients with spinal osteomyelitis have diminished bone density due to bone destruction, and diminished soft-tissue density due to edema and inflammatory exudate. After appropriate antibiotic therapy, follow-up studies reveal an increase in bone density due to healing with osteoblastic activity and new bone formation, and a decrease in size of the soft-tissue mass.[5] CT is also used to guide abscess aspiration and to plan surgical intervention.[2-4]

References

1. Brant-Zawadzki M, Burke VD, Jeffrey RB: CT in the evaluation of spine infection. *Spine* 1983;8:358–364.
2. Burke DR, Brant-Zawadzki M: CT of pyogenic spine infection. *Neuroradiology* 1985;27:131–137.
3. Golimbu C, Firooznia H, Rafii M: CT of osteomyelitis of the spine. *AJR* 1984;142:159–163.
4. Hermann G, Mendelson DS, Cohen BA, et al: Role of computed tomography in the diagnosis of infectious spondylitis. *J Comput Assist Tomogr* 1983;7:961–968.
5. Kattapuram SV, Phillips WC, Boyd R: CT in pyogenic osteomyelitis of the spine. *AJR* 1983;140:1199–1201.
6. Kopecky KK, Gilmor RL, Scott JA, et al: Pitfalls of computed tomography in diagnosis of discitis. *Neuroradiology* 1985;27:57–66.
7. Lardé D, Mathieu D, Frija J, et al: Vertebral osteomyelitis: Disk hypodensity on CT. *AJNR* 1982;3:657–661, *AJR* 1982;139:963–967.
8. Price AC, Allen JH, Eggers FM, et al: Intervertebral disk-space infection: CT changes. *Radiology* 1983;149:725–729.
9. Raptopoulos V, Doherty PW, Goss TP, et al: Acute osteomyelitis: Advantage of white cell scans in early detection. *AJR* 1982;139:1077–1082.
10. Rawlings CE III, Wilkins RH, Gallis HA, et al: Postoperative intervertebral disc space infection. *Neurosurgery* 1983;13:371–375.
11. Sapico FL, Montgomerie JZ: Pyogenic vertebral osteomyelitis: Report of nine cases and review of the literature. *Rev Infect Dis* 1979;1:754–776.
12. Whelan MA, Schonfeld S, Post JD, et al: Computed tomography of nontuberculous spinal infection. *J Comput Assist Tomogr* 1985;9:280–287.

Index

A

Abdominal injuries, spinal injuries and, 246
Achondroplasia, 58, 64-65
Aneurysmal bone cyst, 129
Ankylosing spondylitis, 99, 104-106, 200
　fatty replacement of paraspinal musculature, 104-105
　sacroiliitis, 99
　scalloping of laminae, 104
　spinal injuries resulting from, 105
　spine, characteristics of, 105
Anterior internal vertebral veins, 236, 237, 238
Arachnoiditis, 276-280
Articular facet hypertrophy, 48
Artifacts, 44-46
Astrocytomas, 144, 145, 146
Atlas
　fractures of, 208-210
　pseudospread of, 209
Avulsion of nerve root, 220-223
Axis, fractures of, 212-214

B

Basivertebral veins, 236-238
Bilateral locked facets, 226-228
Biopsy, vertebral, 109
Bulging annulus, 8, 9, 16-18
Bullet injuries, 252-253
Burst fracture, 230-233
Butterfly vertebra, 216-218

C

Calcification
　arachnoiditis and, 278, 279-280

chordoma of the spine, 132-133
ligamentous, 78-79
meningioma, 136-137
scleroderma with, 206
thoracic disc herniation, 42
Case presentations
　achondroplasia, 63-65
　ankylosing spondylitis, 104-106
　arachnoiditis, 275-280
　artifacts, 43-46
　basivertebral veins, 235-238
　bulging annulus, 15-18
　bullet injuries, 251-253
　burst fracture, 229-233
　butterfly vertebra, 215-218
　central disc herniation, 7-10
　central spinal stenosis, 57-62
　cervical disc herniation, 37-40
　cervical spondylosis, 149-150
　Chiari malformation, 193-196
　chordoma of the sacrum, 157-162
　chordoma of the spine, 131-134
　conjoined nerve roots, 27-31
　cranial settling in rheumatoid arthritis, 201-203
　"Cupid's Bow," 111-113
　cystic nerve root sleeve dilatation, 33-36
　degenerative spondylolisthesis, 89-92
　diastematomyelia, 183-186
　dental filling artifact, 43-46
　Ependymoma of the filum terminale, 151-156
　epidural hematoma, 267-269
　epidural metastasis, 115-120
　extruded disc, 19-22
　facet distraction injury, 243-246
　fat graft, 273-274
　fracture of C2, 211-214

fracture, horizontal with diastasis, 239-242
gas, demonstration of, 54-56
hemangioma, 125-126
intramedullary tumor, 143-147
intraosseous gas, 54-55
Jefferson fracture, 207-210
lateral disc herniation, 23-26
lateral recess stenosis, 67-71
lateral thoracic meningocele, 163-166
locked facets, bilateral, 225-228
meningioma, 136-138
metastasis, vertebrae, 107-110
metastasis with intraosseous gas, 53-56
myeloma, 121-124
myelomeningocele, 177-181
nerve root avulsion, 219-223
neural foraminal stenosis, 73-76
neurofibromatosis, 167-170
ossification of the posterior longitudinal ligament, 77-81
osteoarthritis of facet joints, 47-50
osteoblastoma, 127-130
posterolateral disc herniation, 1-6
postoperative free disc fragment, 255-258
postoperative fibrotic scar, 259-261
pseudoherniation of the disc, 11-14
pseudomeningocele, 271-272
pyogenic osteomyelitis, 281-285
pyogenic sacroiliitis, 97-102
rheumatoid arthritis with C1-C2 subluxation, 197-200
sacral fracture, 247-250
schwannomas, multiple intradural, 139-141

scleroderma with massive calcification, 205–206
Smith fracture, 239–242
spondylolysis, 83–88
surgical fusion, 263–266
synovial cyst, 51–52
syringomyelia, 171–175
tethered cord with lipomatous infiltration, 187–192
thoracic disc herniation, 41–42
tuberculous osteomyelitis, 93–96
Cauda equina syndrome, 42, 104
Central disc herniation, 8–10
Central spinal stenosis, 58–62, 76
Cervical disc herniation, 38–40, 223
Cervical spondylosis, 40, 150
Chance fracture, 240, 241, 242, 244
Chemonucleolysis, 22
Chiari malformation, 172, 178, 180, 194–196
Chondrosarcoma, 133–134
Chordoma
  sacrum, 158–162
  spine, 132–134
Clefts, congenital, 209
Computed tomography,
  myelography technique, 4–5, 39, 145, 172, 174, 276; See also Myelography; specific disorders
Computed tomography, technique
  conventional, 12–13, 38–39, 48, 58–59, 100, 212, 276
  "highlighting," 28–31, 152
  "with intrathecal contrast, 4–5, 39, 145, 172, 174
  with intravenous contrast, 39
Congenital abnormalities
  aplasia, pedicle, 217–218
  butterfly vertebra, 216–218
  clefts, 209
  os odontoideum, 216–217, 218
Congenital malformation syndromes
  achondroplasia, 58, 64–65
  Down's syndrome, 200
  Marfan's syndrome, 104, 200
Congenital spinal stenosis, 58–59, 64–65
Conjoined nerve roots, 28–31
Contrast-enhanced CT, intrathecal
  See Computed tomography, myelography
Contrast-enhanced CT, intravenous
  of cervical disc herniation, 39
  of fibrotic scar, 257, 260
  of inflammatory mass, 94, 284
  of neoplasm, 116, 145–147, 152–153

technique, 39, 260
Cranial settling in rheumatoid arthritis, 202–203
"Cupid's Bow," 112–113
Cystic nerve root sleeve dilatation, 34–36

D

Degenerative disc disease, gas and, 54–55
Degenerative spinal stenosis, 58–62, 68–71, 74–76
Degenerative spondylolisthesis, 48, 55, 90–92
Dental filling artifact, 44–46
Diastematomyelia, 184–186
"Disappearing lamina," 240–242
Disc herniation
  central disc herniation, 8–10
  cervical disc herniation, 38–40, 150, 223
  classification, 8–9
  clinical features, 8, 24, 26, 28, 39, 42, 62, 257, 258
  compared to fibrotic scar, 256–258
  extruded disc, 8, 9, 20–22, 256–257
  free disc fragment, 8, 9, 20–22, 256–257
  hard disc, 3, 4
  lateral disc herniation, 4–8, 24–26, 38, 39
  lateral recess stenosis, and, 71
  myelography, comparison with CT, 3–6
  posterolateral disc herniation, 2–6, 29, 38, 39
  posttraumatic, 223
  subligamentous disc herniation, 8–9, 20, 22
  thoracic disc herniation, 42
  See also specific types of herniation
Disc, normal CT appearance, 2–3
Down's syndrome, 200
Dysraphism
  Chiari malformation, 178, 180, 194–196
  diastematomyelia, 184, 185
  forms of, 178
  lipomas and, 179
  meningocele, 178
  myelomeningocele, 178–179
  spina bifida, 178
  tethered cord, 178, 188–192

E

Ependymoma, 140, 141, 144, 146, 152–156, 160, 162

Epidural gas, 55
Epidural hematoma, 268–269
Epidural metastasis, 116–120
Ewing's sarcoma, 160
Extruded disc, 8, 9, 20–22
  postoperative, 256–258

F

Facet artifact, 46
Facet distraction, seat belt injury, 244–246
Facets
  bilateral locked facets, 226–228
  dual innervation, 50
  distraction injury, 244–246
  gas, 55
  hypertrophy, 48, 49
  normal appearance, 48
  osteoarthritis, 48–50
  perched, 228
  subluxation, 221, 222
  unilateral locked facets, 226, 227
  vacuum facet, 48, 55
Failed back surgery syndrome, 71, 76
Fat graft, 274
Fibrotic scar, 256–257, 258, 260–261
Flexion compression injury, 240
Fractures
  atlas (C1), 208–210
  axis (C2), 212–214
  burst, 230–233
  Chance, 240–241, 242, 244, 246
  hangman's, 212–213, 214
  horizontal with diastasis, 240–242
  Jefferson, 208–210, 212
  odontoid process, 212, 214
  pillar, 220–221, 222
  sacrum, 248–250
  sagittal, 230–231
  Smith, 240–242, 244, 246
  wedge, 230
  See also Trauma
Free disc fragment, 8, 9, 20, 21, 22, 256–258
Fusion, spinal, 264–266

G

Gas
  CT demonstration of, 54–56
  degenerative disc disease, 54–55
  epidural, 55
  facet joints, 55
  herniated disc, 55
  intraosseous, 54–55

synovial cyst, 52, 55
thecal sac, 55-56
vacuum disc phenomena, 54-55
Giant cell tumors, 160
Gout, sacroiliitis, 99
Gliomas, 144;
See specific tumors

H

Hangman's fracture, 212-213, 214
Hemangioma, 126
Hemangioblastomas, 140, 141
Hemangiopericytoma, 126
Hematoma, epidural, 268-269
Hematoma, intraspinal, 252
Herniated disc, gas and, 55
"Highlighting," 28-31
Horizontal fracture with diastasis, 240-242
Hypertrophy, articular processes, 48, 49

I

Intradural extramedullary tumors, 136-137
Intramedullary tumor, 144-147
Intraosseous gas, metastasis with, 54-55
Isthmic spondylolisthesis, 84-87, 88

J

Jefferson fracture, 208-210, 212

K

Kyphoscoliosis, neurofibromatosis, 168

L

Lateral disc herniation, 4-5, 8, 24-26, 38, 39
differential diagnosis, 24
Lateral recess stenosis, 68-71
Lateral thoracic meningocele, 164-166
Ligamenta flava, hypertrophy, 58, 61
Limbus vertebra, 217
Lipomas, 144, 161, 178, 179
tethered cord with, 188-192
Lipomyelomeningocele, 179
Locked facets, 226-228

M

Magnetic resonance imaging
Chiari malformation, 194, 196
cordoma of the sacrum, 159
dysraphism, 180-181
intramedullary tumor, 147
lipoma, 189, 192
syringomyelia, 147, 175
tethered cord, 189, 192
Marfan's syndrome, 104, 200
Meningioma, 136-138, 140, 141
Meningocele
dysraphism, 178, 179
intrasacral, 160-161, 162
lateral thoracic, 164-166
Metastasis
biopsy, vertebral, 109
epidural, 116-120
intradural, 140
intramedullary, 144, 146
with intraosseous gas, 54-55
sacral, 117, 158-159
vertebral, 108-110, 118-120, 133
Multicentric reticulohistiocytosis, 199-200
Multiple intradural schwannomas, 140-141
Multiple spinal ependymomas, 140
Myelocele, 178;
See also Myelomeningocele
Myelography
accuracy of, 4
bulging annulus, 17
cystic nerve root sleeve dilatation, 34
of disc herniation, comparison with CT, 3-6
diastematomyelia, 185
epidural metastasis, 116
intramedullary tumor, 145
lateral disc herniation, 4-5, 24
meningioma, 136
posterolateral disc herniation, 6
spinal stenosis, 61
Myeloma, 122-124
POEMS syndrome and, 123
Myelomeningocele, 173-181

N

"Naked facet," 228, 244-246
Nerve room avulsion, 220-223
Neural foramen, stenosis of, 74-76
Neurilemoma, 159-160
Neurofibroma, 35, 137, 138, 140, 154, 159, 164;
See also Schwannomas
Neurofibromatosis, 140, 141, 168-170
lateral thoracic meningocele and, 164
multiple intraspinal tumors and, 140, 141
See also Schwannomas

Neurofibrosarcoma, 154, 155

O

Odontoid fractures, 212, 214
Os odontoideum, 216-217, 218
Ossification of the posterior longitudinal ligament, 78-81
Osteitis, radiation, 161
Osteoarthritis of facet joints, 48-50, 55, 58
Osteoarthritis, sacroiliac joint, 102
Osteoid osteoma, 128-129
Osteomyelitis
pyogenic, 282-285
sacrum, 160
tuberculous, 94-96
Osteosarcoma, 133-134, 160, 161

P

Paget's disease, 133, 160, 161
Pedicle, congenital absence of, 217-218
Perched facet, 228
Perineural cysts, 34-36
Pillar fracture, 220, 221, 222
POEMS syndrome, 123
Posterolateral disc herniation, 2-6, 9, 29, 38, 39
nerve root compression, 2-3, 8
Postoperative bone fragment, 258
Postoperative CT evaluation for
arachnoiditis, 286-280
bone fragment, 258
bone graft, 80, 264-266
disc herniation, 256-258
fibrotic scar, 256-257, 260-261
gas, thecal sac, 55-56
osteomyelitis, 282-285
pseudomeningocele, 272
spinal fusion, 264-266
Postoperative fibrotic scar, 256-257, 260-261
Postoperative free disc fragment, 256-258
Postoperative pyogenic osteomyelitis, 282, 283, 284
Pseudoherniation of the disc, 12-14, 88, 91
Pseudomeningocele, 272
Pseudospread of the atlas, 209
Pseudosubluxation, 233
Pyogenic osteomyelitis, 282-285
Pyogenic sacroiliitis, 98-102

R

Radiation osteitis, 161
Radionuclide scanning

metastasis, 108
myeloma, 123
osteoid osteoma, 128
osteomyelitis, 283
Retrospondylolisthesis, 48, 49
Retrovertebral plexus of veins, 236
Rheumatoid arthritis
  anatomical considerations, 198
  C1-C2 subluxation in, 198-200
  cranial settling, 202-203
  spinal cord compression and, 198-200

S

Sacral cysts, 34, 36
Sacral fracture, 248-250
Sacroiliitis, 98-102
  ankylosing spondylitis, 99
  gout, 99
  pyogenic, 98-99
Sacrum
  chordoma of, 158-162
  ependymoma, 160-162
  Ewing's sarcoma, 160
  fracture of, 248-250
  giant cell tumor, 160
  intrasacral meningocele, 161
  intrasacral tumor, 160-162
  metastasis, 117, 158-159
  neurolimoma, 159, 160
  osteomyelitis, 160
  osteosarcoma, 160, 161
  Paget's disease, 160, 161
  technique, CT, 100
Sagittal fracture, 230-231
Scalloping, vertebral
  ankylosing spondylitis, 104
  neurofibromatosis, 168, 169, 170
Schmorl's node, 109
Schwannomas
  multiple intradural, 140-141
  See also Neurofibroma
Scleroderma with massive calcification, 206
Scoliosis, 13, 144, 148
Screw artifact, 46
Seat belt injury, 240-242, 244-246
Shoulder artifact, 46
Smith fracture, 240-242, 244-246
Spina bifida, 178, 179, 180

Spinal stenosis
  central stenosis, 58-62
  classification, 58
  congenital, 59, 62, 64-65
  degenerative, 58, 60-62
  lateral recess, 68-71
  ligamentum flavum, thickening, 58, 61
  neural foramen, 74-76
  surgical fusion, 265-266
  symptoms of, 62, 71
  See also specific types of spinal stenosis
Spondylolisthesis
  degenerative, 48, 55, 90-92
  isthmic, 84, 87, 88
  pseudoherniation of the disc, 13, 88, 91
  retrospondylolisthesis, 48-49
  spondylolysis, 84, 87
Spondylolysis, 84-88
Spondylolysis, cervical, 40, 150
Stenosis
  See Spinal stenosis
Subligamentous disc herniation, 8, 9, 20, 22
Subluxation of facet joints, 48, 49
Surgical fusion, 264-266
Synovial cyst, 52
Syringohydromyelia, 172, 178, 179
Syringomyelia, 172-175, 223

T

Tarlov cyst, 34-36
Technique, computed tomography
  conventional, 12-13, 38-39, 48, 58-59, 212, 276
  "highlighting," 28-31, 152
  intrathecal contrast-enhanced, 4-5, 39, 145, 172, 174, 276
  intravenous contrast-enhanced, 39
Tethered cord, 188-192
Thoracic disc herniation, 42
Trauma
  bullet injuries, 252-253
  "disappearing lamina" sign, 240-242
  facets, distraction injury of, 244-246
  facets, locking of, 226-228

facets, perching of, 228
facets, subluxation of, 221, 222
hematoma, intraspinal, 252
herniated disc, posttraumatic, 223
"naked facet" sign, 228, 244-246
nerve root avulsion, 220-223
sacroiliac joint, diastasis of, 249, 250
seat belt injuries, 240-242, 244-246
spinal cord edema, 223
syringomyelia, posttraumatic, 174-175, 223
uncovertebral joint distraction, 227
See also Fractures
Tuberculous osteomyelitis, 94-96
Tumors
  See specific types of tumors

U

Uncovertebral joint distraction, 227
Unilateral locked facets, 226-228

V

Vacuum disc phenomena, 16, 17, 54-55
Vacuum facet phenomenon, 16, 48, 55
Veins
  anterior internal vertebral veins, 236, 237, 238
  basivertebral, 236-238
  radicular, 236, 238
  retrovertebral plexus of veins, 236
Vertebral metastasis, 108-110, 116, 118-120, 133
Von-Hippel-Lindau disease, 141

W

Wedge fracture, 230